Resource Competition and Community Structure

MONOGRAPHS IN POPULATION BIOLOGY

EDITED BY ROBERT M. MAY

1. The Theory of Island Biogeography, by Robert H. MacArthur and Edward O. Wilson
2. Evolution in Changing Environments: Some Theoretical Explorations, by Richard Levins
3. Adaptive Geometry of Trees, by Henry S. Horn
4. Theoretical Aspects of Population Genetics, by Motoo Kimura and Tomoko Ohta
5. Populations in a Seasonal Environment, by Stephen D. Fretwell
6. Stability and Complexity in Model Ecosystems, by Robert M. May
7. Competition and the Structure of Bird Communities, by Martin Cody
8. Sex and Evolution, by George C. Williams
9. Group Selection in Predator-Prey Communities, by Michael E. Gilpin
10. Geographic Variation, Speciation, and Clines, by John A. Endler
11. Food Webs and Niche Space, by Joel E. Cohen
12. Caste and Ecology in the Social Insects, by George F. Oster and Edward O. Wilson
13. The Dynamics of Arthropod Predator-Prey Systems, by Michael P. Hassell
14. Some Adaptations of Marsh-nesting Blackbirds, by Gordon H. Orians
15. Evolutionary Biology of Parasites, by Peter W. Price
16. Cultural Transmission and Evolution: A Quantitative Approach, by L. L. Cavalli-Sforza and M. W. Feldman
17. Resource Competition and Community Structure, by David Tilman

Resource Competition and Community Structure

DAVID TILMAN

PRINCETON, NEW JERSEY
PRINCETON UNIVERSITY PRESS
1982

Published by Princeton University Press, Princeton, New Jersey
In the United Kingdom: Princeton University Press, Guildford, Surrey

Library of Congress Cataloging in Publication Data will be found on the last printed page of this book
This book has been composed in Monophoto Baskerville
Clothbound editions of Princeton University Press books are printed on acid-free paper, and binding materials are chosen for strength and durability
Printed in the United States of America by Princeton University Press, Princeton, New Jersey

FOR

Cathie

Lisa, Margie, and Sarah

Preface

This book was motivated by several factors. First, a large body of observational and experimental data which relates aspects of plant community structure to resource availabilities has accumulated in ecology at the same time as the theoretical literature has seen numerous papers on resource competition. I felt that it was time to integrate observation and theory. Second, I had been working on several papers which extended resource competition theory to multispecies communities, as well as on a general review of the observational and experimental literature. In considering submitting one of these papers to *The American Naturalist* for review, I was told that there could be a two-year lag between acceptance and publication. (I am now told that this delay has been reduced.) Third, I received a call from Robert May, the editor of this series, asking if I would be interested in writing a monograph, which could also appear in two years. I chose to write a monograph, which has proven to be both a pleasure and a challenge. The pleasure has come from the opportunity to synthesize theory and observation in much greater depth than would be possible in a series of journal articles. However, each attempt at a synthesis raises new questions. These new questions vex, perplex, and delight me. Some speculative solutions are offered in the final two chapters of this book.

This work presents a simple, graphical theory of multispecies competition for resources. I believe it may be used to predict the dependence of the species composition and species richness of communities on the availabilities of limiting resources. My hope is that this book may alert others to the potential power of the resource approach to community structure and thus may encourage research in a sufficiently diverse array of communities that we may come to know the role of resource competition in structuring natural communities.

In the initial stages of planning this book, I had hoped that Stephen Hubbell of the University of Iowa could be a co-author Unfortunately, his schedule did not allow it. I am sure that the breadth and depth of the work have suffered for his absence. It has, however, been greatly improved because of the comments, criticisms, assistance and encouragement of many others. All errors are mine. I would especially like to thank Robert May for his support and encouragement throughout the process of writing. I was deeply touched by the kindness of G. Evelyn Hutchinson who graciously interrupted work on the fourth volume of his treatise to review the manuscript. I am indebted to many others for comments on part or all of this book or on the ideas behind it, including J. David Allan, Edward Cushing, Francis Evans, Peter Kilham, Susan Kilham, Patrice Morrow, Phil Rundel, and Pat Werner. George Sugihara provided extensive comments which have resulted in significant changes, for which I thank him. I also thank an anonymous plant ecologist (apparently British) who reviewed the book for Princeton University Press and who also provided major comments that have led to significant changes. The first group to read the work in progress were eleven University of Minnesota graduate students whose demands for clarity and brevity strengthened it. I thank Robert Askins, Gloria Biermann, Sue Braun, Naomi Detenbeck, Jesse Ford, Becky Goldberg, Billy Goodman, Richard Kiesling, Pat Lewis, David Rugg, and Robert Sterner. I also thank Mark McKone for his comments. I thank Dan Sell, Heath Carney, and Gary Fahnenstiel for bringing to my attention a multiplication error made in an earlier publication (Tilman, 1977), which is corrected in Figure 54. Their comment led me to start reanalyzing some field data at which I had not looked for 6 years. I would also like to thank Diane Davidson and a group of University of Utah graduate students, who provided numerous comments during the final stages of proofreading. I also thank my wife for help with proofreading. The figures were prepared by Sue McEachran, for whose assistance I express deep appreciation. Laura Carlson typed many more drafts than either she or I cares to remember, and meticulously corrected numerous grammatical and typographical errors. I am grateful to my wife, Cathie, who endured some lonely evenings and a husband overly preoccupied with a book—as well as to my children who have lifted my spirits when all else seemed to be going astray. Portions of this book are an expansion of my "Resources: A graphical-mechanistic approach to competition and predation," published in *The American Naturalist* 116:362–393 (1980), and used with the permission of The University of Chicago Press. Figures 17, 85, and 90 are adapted from articles which first appeared in *Science* and *The American Naturalist*, and are used with their permission. Lastly, I thank the Graduate School of the University of Minnesota for the summer research stipend which allowed me to begin the writing, and National Science Foundation grants DEB-77-06487-A01 and DEB-79-04250 for consistent support of the research which generated the ideas developed herein.

Contents

Resource Competition and Community Structure

CHAPTER ONE

Introduction

> In any study of evolutionary ecology, food relations appear as one of the most important aspects of the system of animate nature. There is quite obviously much more to living communities than the raw dictum "eat or be eaten," but in order to understand the higher intricacies of any ecological system, it is most easy to start from this crudely simple point of view.
>
> G. E. Hutchinson (1959)

In his "Homage to Santa Rosalia," G. E. Hutchinson posed what I believe is the most fundamental question that an ecologist can address: Why are there so many kinds of plants and animals? The most basic qualities of a natural community are the kinds and number of species living in them. What could be more central to the study of natural communities than the question of the causes of patterns in organic diversity? Hutchinson suggested that the major determinant of diversity patterns would be the trophic interactions of organisms. This idea has spawned a wealth of experimentation and theory. This premise—that much of the structure of communities can be understood by knowing the feeding relations of species—is the basis of this book. The processes that Hutchinson called "food relations" have since been termed "consumer-resource interactions" and "resource competition."

In this book, the consumer-resource approach to population ecology is developed to the point where it may be called a theory of the structure and functioning of multispecies communities. As a first attempt at such a synthesis, this work focuses on plant communities and deals primarily with the types of broad and general questions which Hutchinson (1959) raised. Communities have many characteristics which can be considered aspects of their structure, including species composition, species diversity, the relative abundances of species, spatial and temporal patterning of species abundances, and morphological characteristics of the dominant species. The emphasis of this study is on species composition and diversity of communities as related to consumer-resource interactions. Admittedly, numerous stochastic processes—from the level of demographic stochasticity (May, 1973), to the chaos of populations growing at discrete time intervals, to fluctuations in various physical parameters in the natural world—decrease the ability of any theory to predict accurately the dynamics of populations. However, this book is based on the assumption that the interactions among species and of species with their environment are sufficiently strong as to establish major patterns which are discernible over such stochastic noise.

The consumer-resource theory developed here is used to explore such questions as:

How may resource competition help determine which species will be dominant in a community and which will be rare?

What limits does resource competition place on the diversity of a community?

How do spatial structure, temporal variability, and trophic complexity influence species diversity?

Is the structure of plant communities inherently different from that of animal communities because of the different types of resources they consume?

What prevents the evolution of a superspecies, i.e., a species which is such a superior competitor for resources that it displaces all other species?

Should the structure of communities of motile animals differ markedly from that of sessile animals?

How may resource competition influence succession?

By no means are all of these questions given equal weight. My main purpose is to present a simple, graphical theory of resource competition and to demonstrate the implications of resource competition for the species composition and diversity of natural communities. Although the theory may apply to both animal and plant communities, most of the discussion concentrates on the relationships between the availabilities of limiting resources and the structure of natural and manipulated plant communities. The final chapters provide some speculations concerning resources and animal communities and briefly explore some of the other questions raised above.

This book is a blend of theoretical, experimental, and correlational information. The experimental and correlational information comes from numerous studies of both terrestrial and aquatic plant communities. The theory presented is a graphical approach to the equilibrium structure of communities and their resources. The graphical theory is an extension of a body of work on consumer-resource interactions based on papers such as Rapport (1971), MacArthur (1972), Covich (1972), Maguire (1973), Stewart and Levin (1973), Leon and Tumpson (1975), Petersen (1975), Taylor and Williams (1975), Rapport and Turner (1975, 1977), Abrosov (1975), and Tilman (1977, 1980). An equilibrium approach is used because it explores the long-term effects of the consumer-resource process and because an assumption of equilibrium is inherently simpler than an assumption that a given observation can only be understood in terms of non-equilibrium processes. This is not to imply that non-equilibrium processes are unimportant, but rather that it is most appropriate to consider the simpler, equilibrium explanations first. A graphical approach to theory is used for several reasons. First, graphical theory can be a powerful and general technique, allowing questions to be explored without the necessity of describing cases with specific equations. Second, equations may hide assumptions (at times biologically unrealistic ones) which may be laid bare in a graph. Third, many of the ideas presented in this book were formed using graphical techniques, because I often think in terms of graphical relationships. Lastly, I believe that graphs can communicate ideas more effectively and to a broader audience than can a series of equations.

This work uses a broad definition of resources. Just as nitrate, phosphate, and light may be resources for a plant, so may nectar, pollen, and a hole in a log be resources for a bee, and so may acorns, walnuts, other seeds, and a larger hole in a log be resources for a squirrel. All things consumed by a species are potentially limiting resources for it. Consumers may compete for prey and, in turn, be the point of competition for their predators. When extended to several trophic levels, this approach uses information on the resource requirements of several species to predict the outcome of interactions among them.

The approach developed here, which emphasizes the mechanisms of interactions among species, may be contrasted with the more phenomenological and descriptive approach of classical theory. The Lotka-Volterra competition equations do not explicitly consider the mechanism of interaction among species, but instead summarize these interactions in a few parameters, the competition coefficients. This has made it difficult to know how different mechanisms of interaction determine the values of the competition coefficients. As is shown in Chapter 7, the values of the Lotka-Volterra competition coefficients depend on the type of resource, the consumption characteristics of the species, and the processes governing the supply of the resources. Thus, the values of the competition coefficients can only be theoretically forecast when an explicit, mechanistic model of competition is used, such as was done by MacArthur (1972). Once such a model is formulated, it seems simpler to use the mechanistic model itself, rather than to use it to estimate parameters which are put back into the more descriptive Lotka-Volterra model.

In addition, the apparently simpler classical equations of competition are not always simpler in actuality. For instance, if the Lotka-Volterra equations were used to describe competition between two species, it would be necessary to estimate values for six constants ($\alpha, \beta, r_1, r_2, K_1, K_2$). A mechanistic approach, such as in Tilman (1977) would also require six constants ($k_1, k_2, r_1, r_2, Y_1, Y_2$). If, however, competition between 10 species were to be considered, the Lotka-Volterra equations would require the estimation of 90 pairwise alphas and 10 pairs of K_i and r_i, for a total of 110 parameters. In contrast, a resource-based approach would only require three parameters per species, for a total of 30 parameters. Thus, the resource-based model may be simpler when applied to many species.

There are many correlational, observational, and experimental data sets which suggest that resources are an important factor structuring natural communities. For instance, the Rothamsted experiments, started in 1856 by Lawes and Gilbert and still in progress today, have shown a dramatic effect of fertilization on the structure of plant communities. Heavily fertilized plots have had species richness decline from about 40 species to about 3 or 4 species, whereas unfertilized plots retain their original diversity. The type of fertilization (i.e., the ratios of various mineral elements in the fertilizer) strongly influences which plant species dominate plots. These experiments, the most long-term investigation yet performed in ecology, are discussed in considerable detail in Chapters 5 and 6. Similar studies performed for both terrestrial and aquatic plant communities have also demonstrated the effects of resource enrichment and nutrient ratios on community structure. Disturbance is a process which provides open sites for sessile animals and light for terrestrial plant seedlings. The studies of Paine (1966, 1969), Harper (1969), Dayton (1971), Grubb (1977), Lubchenco (1978), Connell (1978), Platt and Weis (1977), and others have indicated that disturbance rate, which may be a measure of the rate of supply of limiting resources, strongly influences the species composition and diversity of communities of intertidal animals and plants, coral reefs, tropical forests, prairies, and communities of sessile organisms in general. Such studies are discussed is relation to a simple, graphical theory of competition for open space, which is developed in Chapter 8.

Many of the trends reported in plant and animal communities seem to be at odds with each other as to the role of resources in determining the diversity of the communities. For instance, the field work done to date in plant communities has indicated that at most three or four resources are limiting in any community. In terms of many of the theories of resource competition, it seems hard to imagine how hundreds of plant species may coexist when limited by a few resources, all of which are required for plant growth. On the grossest global level, plant communities may have 30 different limiting resources, and yet there are more than 300,000 species of terrestrial plants, giving approximately 10^4 plant species globally for each limiting resource. In contrast, terrestrial herbivorous animals may have as their limiting resources all the various parts of these 300,000 species of plants, and there are at most one million species of such animals. Thus, there are approximately 10^4 plant species for every limiting plant resource but a few animal species for every animal resource. Can this difference be understood in terms of the different types of resources used by motile animals and plants? Chapter 9 explores this and several other equally broad questions.

In summary, this book addresses the question of the role of resources in structuring natural communities. I start by classifying resources as to type (essential, substitutable, etc.). I then use a simple graphical model to predict foraging strategies for different types of resources (Chapter 2). A simple theory of competition among several species for one limiting resource is developed in Chapter 3. Chapter 4 expands this to a treatment of consumer-resource interactions for two resources, using a graphical approach based on resource-dependent isoclines. Chapter 5 expands this into an equilibrium theory of multispecies consumer-resource interactions in spatially heterogeneous environments, exploring the effects of habitat resource richness on the species diversity of plant communities. Chapter 6 explores the empirical evidence which relates various aspects of plant community structure to resource availability and the ratios of supply of limiting resources. The interrelationships between classical Lotka-Volterra theory and resource competition theory are presented in Chapter 7. The final two chapters are more speculative. Chapter 8 considers cases of competition for space and food in communities of sessile animals, and discusses some general effects of disturbance on community structure. Chapter 9 briefly explores a series of other related questions, and suggests further research which needs to be done to broaden our understanding of the effects of resources on the structure of natural communities.

This book progresses from a fairly simple treatment of the mechanisms of resource competition among a few species to a series of increasingly broad questions about resource competition and community structure. Throughout the text, theory is presented graphically. Mathematical detail is kept to a minimum, with interested readers referred to relevant papers and to the Appendix. Although the theory developed here may apply to both animal and plant communities, the examples discussed are mainly from plant communities. This reflects both the earlier interest of plant ecologists in the role of resources and the relative ease with which plant nutrient resources can be measured compared to many animal resources.

The theory of resource competition and community structure is offered as a testable—and hence potentially falsifiable—theory. The experimental and observational work reviewed in these chapters often seems strikingly consistent with the theory that is developed. However, to date no single community has been studied in sufficient depth to allow a complete test of the possible validity of the theory. If this book accomplishes nothing else, I hope that it will encourage a much more detailed look at the relationships between resources and the structure of both plant and animal communities. The mechanistic basis of the theory allows it to make testable predictions both from lower to higher levels of community organization and from higher to lower levels of organization. For instance, information on the resource requirements of individual species and the availabilities of these resources in a habitat can be used to make testable predictions about which and how many species should coexist in that community. Conversely, the patterns of distribution and abundance of species in a community in relation to resources may be used to make testable predictions about the resource requirements of individual species. Either of these pieces of information could be used to make testable predictions of the effects of various patterns of resource enrichment on community structure. Thus, this theory is capable of making multiple, testable predictions.

There is one obvious but extremely important restriction placed on this theoretical work. For any given community, this theory only applies to limiting resources—to those resources which can be experimentally shown to be limiting the reproduction of at least one member of the community. If this restriction were to be ignored, it would be possible to support or refute the theory independent of its validity. Spurious correlations between community structure and non-limiting resources are sure to exist. Such correlations may be just as likely to seem to support as to refute this theory.

CHAPTER TWO

What Are Resources?

Part of the breadth of the consumer-resource approach which is developed in this book comes from a broad definition of what a resource is. I consider a resource to be any substance or factor which can lead to increased growth rates as its availability in the environment is increased, and which is consumed by an organism. For instance, the growth rate of a plant (often measured as the rate of weight gain or as the rate of seed set, but most appropriately as the long-term instantaneous rate of change once the population has reached a stable age or stage distribution [Hubbell and Werner, 1979]) may be increased by the addition of nitrate, which is consumed by the plant. There may be concentrations of nitrate at which the addition of more nitrate will lead to decreased growth and death. Still, by the definition offered above, nitrate is a resource—a consumable factor which can potentially limit the growth rate of the population. For sessile animals, space (open sites) may be a resource. Increases in the amount of open space can cause increased reproductive rates, and the animals "consume" the open sites as they colonize and grow on them.

Some factors are not resources by this definition. For instance, temperature is not a resource. The reproductive rate of a species may increase with increases in temperature, through some range, but the species does not consume temperature. This is not to imply that temperature and other non-consumable physical and biological factors are not important, but that they must be considered in a different way from resources. It is through depression of resource levels caused by consumption that species may compete with each other, and thus that resources may influence the structure of communities. In a spatially patchy environment, species may compete for hotter or colder spots, examples being the competition that may occur between various lizards for basking sites, or between various species for physically favorable microsites. If this occurs, such differences between sites may be considered to be ranges of a resource, for individuals "consume" such sites, making them unavailable for others.

This verbal discussion of what a resource is may be illustrated graphically. Figure 1 shows several hypothetical and actual resource-dependent population growth curves. Although there may be a variety of shapes for such "growth curves," a factor is a resource only if a graph of per capita reproductive rate against resource availability has a region of increase, and if, for resource availability in that region, the species is consuming the resource, i.e., tending to reduce its availability.

This definition of a resource becomes difficult to use when a species is potentially limited by several factors. For instance, there may be ranges of availability of one potentially limiting factor for which a second potentially limiting factor would be a resource by the definition offered above, and other ranges of availability of the first potentially limiting factor for which the second factor would not be a resource. To overcome this difficulty, let us say that a factor will be considered a resource if there is some range of availability of other limiting factors for which the factor meets the definition offered above. This suggests that the way in which several resources interact to determine the growth rate of a species may be used as the basis for a classification of resource types.

To make such a classification, consider some general equations that express how resources influence the growth of a population and how the consumers affect the availability of the resources (Eq. 1.1). For any quantity, the net rate of change depends on the difference between the rate of supply (birth) of that item and the rate of consumption (death) of that item. The

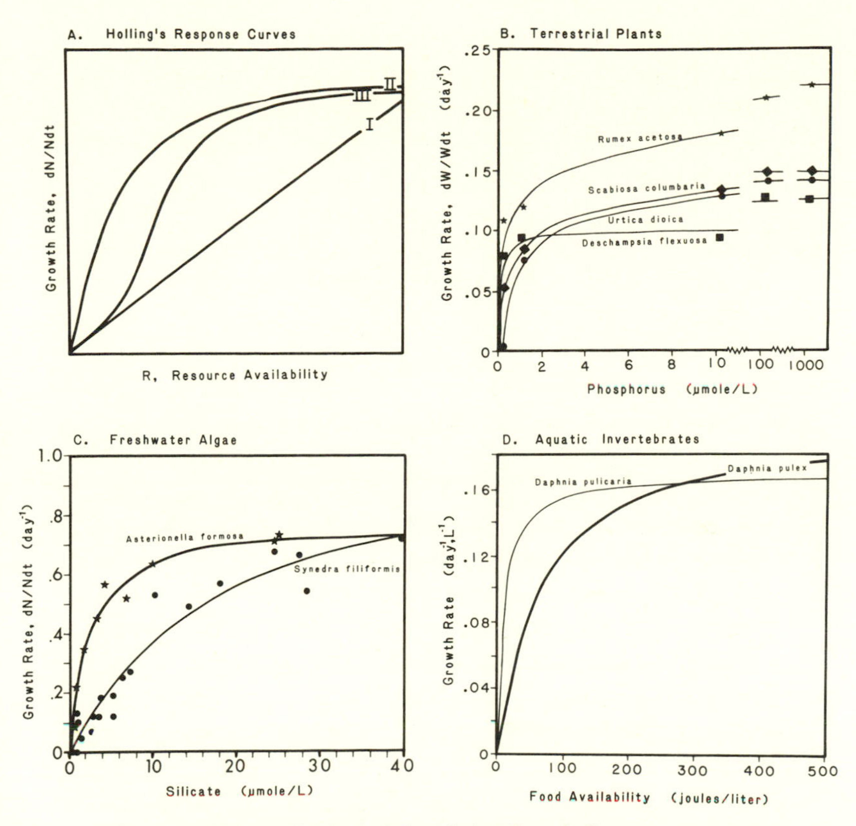

FIGURE 1. Resource-Dependent Population Growth Curves.

A. Three hypothetical curves showing the possible dependence of per capita reproductive rate on resource availability, modified from Holling (1959).

B. Experimentally determined dependence of the rate of weight gain of four terrestrial plants on the availability of phosphate (Rorison, 1968). These curves are hand-drawn through data points showing the estimated specific rate of weight gain for days 10 to 40 at each nutrient treatment employed by Rorison. Note the apparent superior phosphate competitive ability of *Rumex acetosa* over the other three species studied.

C. Silicate-dependent per capita reproduction in two species of freshwater diatoms, *Asterionella formosa* and *Synedra filiformis*, isolated from Lake Michigan (Tilman, 1981). Data have been fitted by a Monod (hyperbolic) curve.

D. The observed dependence of the growth rate of two species of *Daphnia* ("water fleas") on algal density. The two curves shown are the Monod curves which Hrbackova and Hrbacek (1978) fitted to their data.

per individual rate of change of a species, $dN/dt \cdot 1/N$ (which I will write as $dN/N\,dt$ for the remainder of this book) thus depends on the difference between birth and death processes. The rate of change of a resource depends on the difference between supply and consumption rates:

$$\begin{aligned} dN/N\,dt &= \text{reproductive rate} - \text{death rate} \\ dR/dt &= \text{supply rate} - \text{consumption rate.} \end{aligned} \tag{1}$$

Expressed more formally, these may be written as generalized resource competition equations,

$$\begin{aligned} dN_i/N_i\,dt &= f_i(R_1, R_2, \ldots, R_k) - m_i, \\ dR_j/dt &= g_j(R_j) - \sum_{i=1}^{n} N_i f_i(R_1, \ldots, R_k) h_{ij}(R_1, \ldots, R_k), \end{aligned} \tag{2}$$

where N_i is the population density of species i; R_j is the availability of resource j; m_i is the mortality rate of species i; f_i is the function describing the dependence of the net per capita reproductive rate on resource availability; g_j is the function describing the supply rate of resource j; h_{ij} is the function describing the amount of resource j required to produce each new individual of species i; all for a total of n species competing for k resources.

These equations make several important assumptions. First, they assume that species interact only through their use of resources. The equations do not include any direct effect of change in the density of one species on the growth rate of another species. Each species affects all other species and each member of a species affects all other members of the same species and all other members of all other species *only* through its effect on the availability of the limiting resources. This may be stated as $\partial f_i/\partial N_j = 0$ for all i and j. That is, when all other factors are held constant, a change in the population size of species j has no effect on the growth rate of species i, for any values of i and j. A second assumption inherent in these equations is that the resources are not interactive. A change in the availability of one resource is assumed to have no effect on the rate of supply of any other resource. This may be expressed as $\partial g_i/\partial R_j = 0$ for $i \neq j$. This is comparable to saying that the resources are not themselves interacting entities. If resources did interact (see Lynch, 1978), it would be necessary to modify the equations above to include terms for the mechanism of interaction. This could be done in several ways. For instance, if the resources were plants being consumed by animals, the model could be extended explicity to include competition among the plants, thus adding another trophic level to the model. The assumptions that species interact only through use of limiting resources and that resources are non-interactive are two of the more important but less obvious assumptions of this model. The generalized resource competition equations embody many other more obvious assumptions, including continuous reproduction, non-integer individuals, and homogeneous populations and habitats.

The classification of resource types is best illustrated graphically using resource-dependent population growth isoclines. Before discussing such growth isoclines, it is important to define what I mean by growth rate. For organisms with simple life histories, the per capita rate of population growth can easily be estimated with a few censuses of population density, and calculated as $dN/N\,dt = \ln(N_{t+1}/N_t)$. This is the method often used for single-celled organisms such as algae and bacteria, and one which is suitable for any population which has reached a stable age or stage distribution (Vandermeer, 1975). However, I will often use resource-dependent isoclines to refer to the resource requirements of organisms with complex life histories, such as higher plants and animals. For many animals, the per capita rate of reproduction can be estimated using information on the age dependence of survivorship and reproduction, such as is included in the Leslie matrix (Leslie, 1948). For such populations, I am referring to the rate of population change at a stable age distribution when I mention the population growth rate. Plants, in general, have much more complex life histories than animals (Harper, 1977). This greatly increases the difficulties associated with studies of plant populations. As Harper (1977) has discussed, a single individual plant can often have traits which are similar to the traits associated with an animal population. Plant populations, though difficult to study, do have growth rates, just as do all populations. Extending a technique presented by Lewis (1972), Hubbell and Werner (1979) demonstrated that the long-term, per capita rate of population change could be calculated for plant populations with even the most heterogeneous life histories. Hubbell and Werner used λ to symbolize the finite (discrete) rate of population increase which would be reached when the population achieved a stable proportion of all forms of its life history (a stable stage distribution). In this book, when I refer to the growth rate of higher plants I mean the per capita rate of population change when a stable stage distribution has been reached. This is $dN/N\,dt = \ln \lambda$. Thus, for all organisms considered here, I will consider the effects of resources on their average, long-term rate of population increase. This method ignores many potentially interesting consequences of life-history variation. It would be beneficial (but outside the scope of this study) to consider models of resource-dependent competition in which different life history stages were included.

Given this definition of growth rate, it is possible to use resource-dependent growth isoclines to represent the resource requirements of a species. Such "resource growth isoclines" (Tilman, 1980) show all the combinations of resource availabilities for which a given species has a given value for $dN/N\,dt$. Let us first assume that a species is living in a habitat in which there is no mortality, i.e., in which $m_i = 0$, and in which there are two potentially limiting resources. Figure 2 shows several of the shapes that such isoclines could take.

Figure 2A-D shows four different types of resources which are substitutable for each other. These resources are said to be

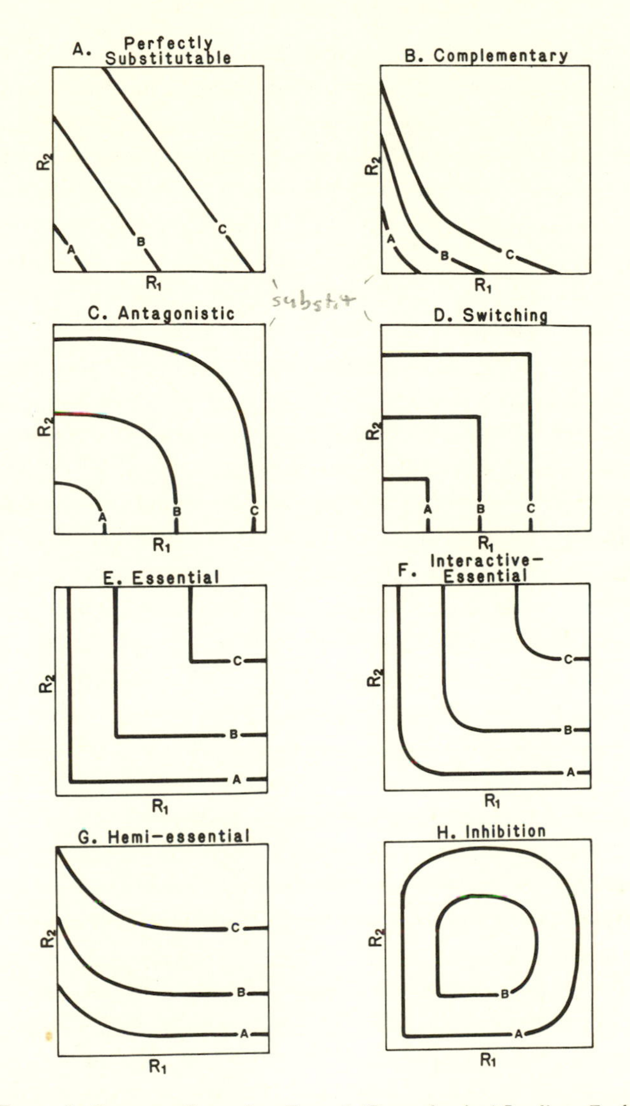

FIGURE 2. Resource-Dependent Growth (Reproduction) Isoclines. Each of the growth isoclines shown above represents the amounts of two resources which would have to exist in a habitat for an individual to have a given reproductive rate. The isoclines thus show the combinations of the concentrations of two resources which lead to a particular reproductive rate. Because reproductive rate increases with resource availability, isoclines further from the origin represent higher reproductive rates. Thus, for all cases shown above, isocline *A* has the lowest reproductive rate, isocline *B* has an intermediate rate, and isocline *C* has the highest reproductive rate.

substitutable for each other in that each can sustain reproduction when the other is lacking. This can be seen in Figure 2A-D by noting that each of these isoclines intersects the R_1 and R_2 axes. This means that a given reproductive rate can be maintained in a habitat which has only one or the other resource, or in a habitat with various intermediate availabilities of the two resources. For perfectly substitutable resources (Fig. 2A), the isoclines are straight lines. One resource can be substituted in direct proportion for the other. The effect of a given amount of R_1 on the growth of this species is comparable to the effect of a constant proportion of this amount of R_2. The three different isoclines of Figure 2A show that the per individual reproductive rate of this species increases with resource availability. Substitutable resources are defined as being complementary if growth isoclines bow in toward the origin (Fig. 2B). Inward-bowing growth isoclines mean that a species requires less of two resources when consumed together than would be estimated by a proportionality constant calculated from the points of intersection with the R_1 and R_2 axes. Two resources are defined as being antagonistic if the growth isoclines bow outward (Fig. 2C). This indicates that a species requires proportionately more resource to maintain a given reproductive rate when two resources are consumed together than when they are consumed separately. The final class of substitutable resources are perfectly antagonistic resources, also called switching resources. The isoclines for switching resources form a right-angle corner pointing away from the origin, as shown in Figure 2D. As will be discussed later in this chapter, such isoclines imply that a species consumes either one or the other resource, but does not consume both at the same time.

Figures 2E and F show isoclines for resources which are essential with respect to each other. Essential resources are required for growth. One essential resource is unable to substitute completely for another essential resource which is in low availability. This can be seen by noting that the isoclines run parallel to both axes. This means that an increase in either one or the other resource will eventually lead to no further increase in growth rate even though the growth rate reached is less than the maximal rate. Figures 2E and F show two varieties of essential resources. The isoclines of Figure 2E form a right-angle corner, indicating that a species is limited by one or the other resource, with absolutely no substitutability of one resource for the other over even a small range of availabilities. Such resources are called perfectly essential. The isoclines of Figure 2F show a curved corner at the point where a species is going from being limited by one to being limited by the other resource. Such curved isoclines imply that there is a range of resource availabilities through which one resource may partially substitute for the other, i.e., through which a lack of one resource may be partially compensated for by a larger amount of the other. Resources with isoclines such as those shown in Figure 2F are termed interactive essential resources.

The final class of resources is termed hemi-essential. For hemi-essential resources (Fig. 2G), growth isoclines cross one of the axes, but eventually run parallel to the other axis. Hemi-essential resources are a pair of resources of which one is required for growth and the other may partially substitute for the first. Growth isoclines for hemi-essential resources intersect the axis of the required resource, and do not cross the axis of the resource which can partially substitute for it.

The isoclines of Figure 2A-G do not show resource inhibition of growth at high availabilities of resources. There are cases in which inhibition does occur. For instance, light, a resource required by most plants, leads to increased growth rates through a broad range of intensities, but can inhibit growth at very high intensities. A plant may be similarly inhibited by high (toxic) availabilities of a nutrient resource. A more complete representation of growth isoclines would include growth isoclines which could form closed curves, as shown in Figure 2H. These curves become closed because growth decreases with increases in the levels of the resources at very high levels. Maguire (1973) considered isoclines which form closed curves because of inhibition, but I will not consider this possibility further in this book because most natural communities have sufficiently low levels of resources that inhibition seems unlikely.

The classes of substitutable, essential, and hemi-essential resources include all the major ways in which a species may respond to two limiting resources. Before using these isoclines to explore competition for resources, it may be useful to discuss the biological basis for these resources classes, and to explore under what circumstances and for what organisms various resources are expected to fall into each class. In doing so, I should note that the definitions just offered are limited to pairwise comparisons of resources. A group of three resources may be perfectly substitutable one with respect to the other, a determination of which could be made by pairwise comparisons. However, such pairwise comparisons do not mean that there would be no complementary or antagonistic effect when the interactions among all three were considered. That there are straight-line isoclines for R_1 and R_2 given that $R_3 = 0$ does not necessarily mean that the three-dimensional resource isocline would form a planar surface. It could bow inward or outward in the middle of the plane and still have straight-line edges. However, as will be shown later, such complications have a small impact on the structure of communities compared to the difference between essential and substitutable resources.

Resource class is not just a property of the nutritional content of the resource, but is the combined result of the nutritional value of the resources, the costs associated with obtaining the resources, and the foraging methods of a species. The foraging methods of a species are a complex response to numerous factors, many of which are not related to nutritional status of the resource. For instance, foraging may be spatially and temporally timed to minimize exposure to predators, and this may increase the consumption of a resource compared to predictions made on a nutritional basis alone. Similarly, two resources may be located in patches some distance from one another so that, even if they were perfectly substitutable with respect to each other nutritionally, they would not be consumed at the same time. If a forager tended to prefer the patches that led to the greatest potential net reproduction, it would then be responding to two nutritionally substitutable resources in a switching manner. Thus, the shape and position of the growth isocline of a species reflect nutritional, behavioral, spatial, and other factors. It is not the purpose of this book to rigorously explore under what circumstances particular isocline shapes should occur, although this is an important question in its own right. Rather, this study demonstrates, for any types of resources, how the outcome of interspecific competition for resources may be predicted. In order to predict the outcomes of resource competition, the shape and position of the resource growth isoclines must be known, as well as the rates of consumption of the various resources. The following discussion will briefly explore the factors that may lead to a given shaped isocline and what the consumption characteristics of species with such isoclines may be.

Resource items that can each support growth on their own are substitutable relative to each other. Many of the dietary items of carnivores are probably substitutable, as may be most of the items consumed by herbivores. Substitutability only defines the end points of the resource growth isoclines—stating that all growth isoclines must touch both the R_1 and R_2 axes. The shape of the curve joining these two points is the basis for the definition of types of substitutable resources.

Perfectly Substitutable Resources

Two resources are perfectly nutritionally substitutable if the two can be substituted for each other with equal effect at all abundances of the two resources. This would mean that consuming an amount of resource 2, R_2, would be equivalent to consuming an amount of C_1R_1 of resource 1, for all values of R_2, with C_1 a constant (i.e., $R_2 = C_1R_1$ and $R_1 = R_2/C_1$). Because of the nutritional complexity of foods, it seems unlikely that many foods would be perfectly substitutable for each other. However, this may be a useful approximation.

There are many different ways in which two nutritionally perfectly substitutable resources may be consumed. An individual's foraging strategy would depend on the patterns of availability of the resources in the habitat, the capabilities of the individual to locate and capture the resources, and other factors. A full treatment of optimal foraging is outside the realm of this book, but a graphical theory of optimal foraging developed by Rapport (1971) and Covich (1972) allows many questions of optimal diet to be easily explored. This graphical theory considers two constraints on diet: consumption or foraging constraints, and nutritional constraints.

The first constraint is that placed on the amounts of two resources which an individual is capable of consuming in a given time period. This constraint comes from the time required for searching, handling and processing food items, the tradeoff inherent in an organism putting its time and energy into consuming one resource and thus having less time and energy for consuming another resource, and also from the temporal and spatial patterning of resource availability in the habitat. This constraint may be illustrated graphically, using consumption constraint curves (Fig. 3). A consumption curve shows the maximal amounts of R_1 and R_2 that an individual is capable of consuming in a given time period. Note that the axes of Figure 3 are the amounts of R_1 and R_2 *consumed* per unit time, and *not* the availabilities of R_1 and R_2. Figures 3A, B, and C show three different shapes for consumption constraint curves, reflecting three qualitatively different ways in which foraging may be constrained. The linear (straight-line) consumption constraint curve of Figure 3A suggests a simple tradeoff between consumption of one or the other resource, such as might occur

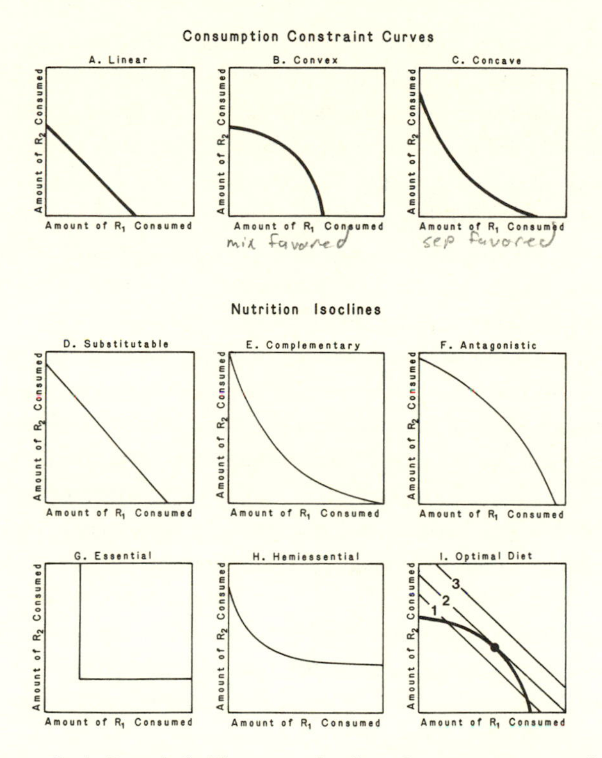

FIGURE 3. A, B, and C. The curves in these figures are consumption constraint curves which show the tradeoff in consumption of two resources in a limited period of time. Each curve defines the maximal amounts of the two resources which an individual is capable of consuming in a given habitat in a given time. The shape and position of the consumption constraint curve an individual experiences will depend on the availabilities of the resources in the habitat, the structure of the habitat, and the foraging methods used by the individual.

D to H. The nutrition isoclines in these five figures show the dependence of per capita reproductive rate on the amounts of the two resources which an individual *consumes*. They show the direct nutritional effects of two resources on reproductive rate, and thus differ in an important way from the growth isoclines of Figure 2. Growth isoclines show the dependence of reproductive rate on resource availability in the habitat, not on the amounts of the resources which are actually consumed.

I. The thick line of this figure is a consumption constraint curve, and the thin lines are nutrition isoclines. The optimal diet for this individual in this habitat is shown with a dot. This diet is the point of tangency between the consumption constraint curve and the nutrition isocline with the highest reproductive rate. Individuals which consume these two resources in the quantitites at the point of tangency will have the highest possible reproductive rate for this habitat.

for two resources which could be obtained using the same foraging method and which were uniformly distributed in a habitat. A linear consumption constraint curve implies that the total amount of food that can be consumed is restricted, but that one resource may be substituted for the other in a simple manner. A convex consumption constraint curve (Fig. 3B) states that a mixed diet is more easily obtained than a diet of either resource alone. This could result from the two resources often occurring together in scattered patches, such that a consumer would easily obtain both resources at the same time but less easily obtain an equivalent amount of one or the other resource. A concave consumption constraint curve (Fig. 3C) means that a consumer faces a cost in consuming intermediate quantities of two resources compared to a diet biased more toward only R_1 or only R_2. The concave curve could be caused by a tendency for resources to occur in single-resource patches, or by a different foraging method being required to obtain each resource. Because there is no *a priori* reason to consider any one of these shapes for the consumption constraint curve sufficiently general that it alone may be used in an analysis of optimal foraging, all will be used in the analysis that follows.

The second constraint needed to predict optimal diet is the relationship between the amounts of two resources which an individual consumes and the reproductive rate of the individual. This is the nutritional dependence of reproduction on consumption, and can be represented by a series of nutrition isoclines (Fig. 3D, E, F, G, H). A nutrition isocline shows the various combinations of R_1 and R_2 that can be consumed to give a certain reproductive rate. Nutrition isoclines for greater reproductive rates are further from the origin, indicating that reproductive rate increases with consumption. Nutrition isoclines differ from resource-dependent growth isoclines in that resource growth isoclines show reproductive rate in relation to *environmental availabilities* of resources, whereas nutrition isoclines show reproduction in relation to the *amounts of resources consumed.* This difference must be kept in mind when studying Figures 3, 4, 6, 7, 8, and 9. The five nutrition isoclines of Figure 3D-H illustrate the five nutritional categories considered in this book.

Given the joint constraints of nutrition isoclines and the consumption constraint curve, it is a simple matter to determine what the optimal diet of an individual should be. The optimal diet would be that point on the consumption constraint curve which leads to the greatest reproductive rate. This is the point of contact between the consumption constraint curve and the nutrition isocline which is furthest from the origin *but still touches the consumption constraint curve.* This is illustrated in Figure 3I for a convex consumption constraint curve and perfectly substitutable nutrition isoclines. The diet is limited to some point on the consumption constraint curve. Of all these points, the point shown with a large dot gives the maximal reproductive rate, because this is the point of intersection of the consumption constraint curve with the nutrition isocline that leads to the highest reproductive rate. The large dot shows the optimal diet, i.e., the optimal amounts of R_1 and R_2 to be consumed. The dot is at the point of tangency of nutrition isocline 2 with the consumption constraint curve. The two points of intersection of nutrition isocline 1 with the consumption constraint curve are a suboptimal diet, and nutrition isocline 3 does not intersect the consumption constraint curve, indicating that diets which lead to that reproductive rate are impossible.

For any given shape of the consumption constraint curve, three different cases will be considered (labeled Case 1, Case 2, and Case 3 in Figure 4). For Case 1, the consumption constraint curve reflects relatively low availability of R_1 and relatively high availability of R_2 (Fig. 4, Case 1). The consumption constraint curve of Case 2 reflects approximately equal availabilities of R_1 and R_2. A high availability of R_1 compared to the availability of R_2 gives the consumption constraint curve

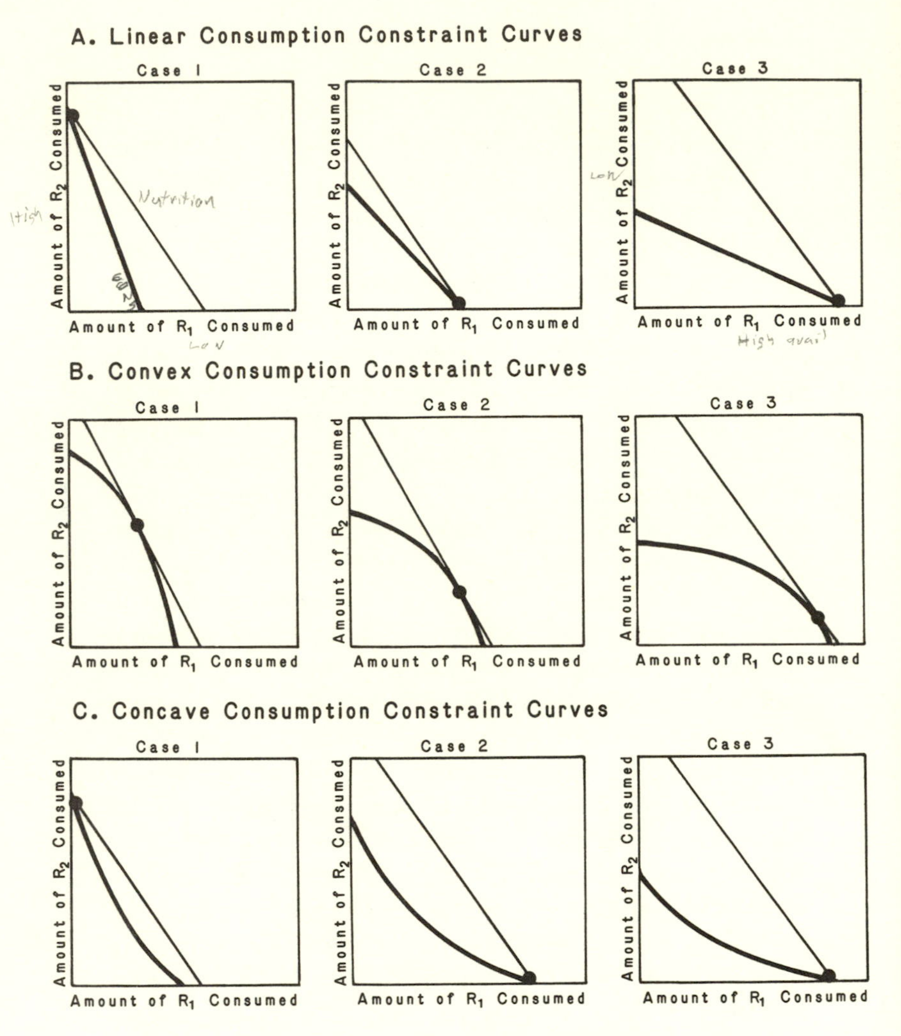

FIGURE 4. Optimal diet for nutritionally perfectly substitutable resources. Part A of this figure shows results for linear consumption constraint curves, part B shows results for convex consumption constraint curves, and part C for concave consumption constraint curves. The three cases shown in each part represent situations with low availability of R_1 (Case 1), low availability of R_2 (Case 3) and intermediate availabilities of the two resources (Case 2). Note that the optimal diet is predicted to switch from only R_2 to only R_1 for increasing availability of R_1 for parts A and C, but that a mixed diet is always optimal for part B.

of Case 3. These three cases allow a qualitative exploration of the dependence of optimal diet on the relative availability of two resources.

The optimal diet for nutritionally perfectly substitutable resources is highly dependent on the shape of the consumption constraint curve and the availability of the two resources. When there are linear consumption constraint curves, the predicted optimal diet consists of only R_2 when R_1 is in low availability (Fig. 4A, Case 1), and of only R_1 when R_2 is in low availability (Fig. 4A, Case 3). There is one proportion of the availabilities of R_1 and R_2 at which all proportions of R_1 and R_2 are an equally optimal diet. Below this proportion, the diet should consist of only R_2, and above it the diet should consist of only R_1. The graphs of Figure 4A predict that a consumer should specialize on the one resource which leads to the greater reproductive rate, and switch to consuming the other resource only when it leads to the greater reproductive rate. A diet consisting of both resource items, i.e., a mixed diet, is never predicted to be optimal for nutritionally perfectly substitutable resources with a linear consumption constraint curve.

The same prediction is made for nutritionally perfectly substitutable resources when there is a concave consumption constraint curve, as shown in Figure 4C. The optimal diet consists of only R_1 or only R_2. A diet consisting of both R_1 and R_2—a mixed diet—is predicted to occur only when there is a convex consumption constraint curve. With a convex curve, nutritionally perfectly substitutable resources are optimally consumed in a mixed diet, as shown in Figure 4B. With such a convex curve, the proportion of a resource in the diet increases as its proportionate availability increases.

Nutritionally perfectly substitutable resources consumed as shown in Figure 4B give linear resource-dependent growth isoclines and consumption vectors approximately as shown in Figure 5A. The consumption vectors of Figure 5A illustrate the relative rates of consumption of R_1 and R_2 per individual

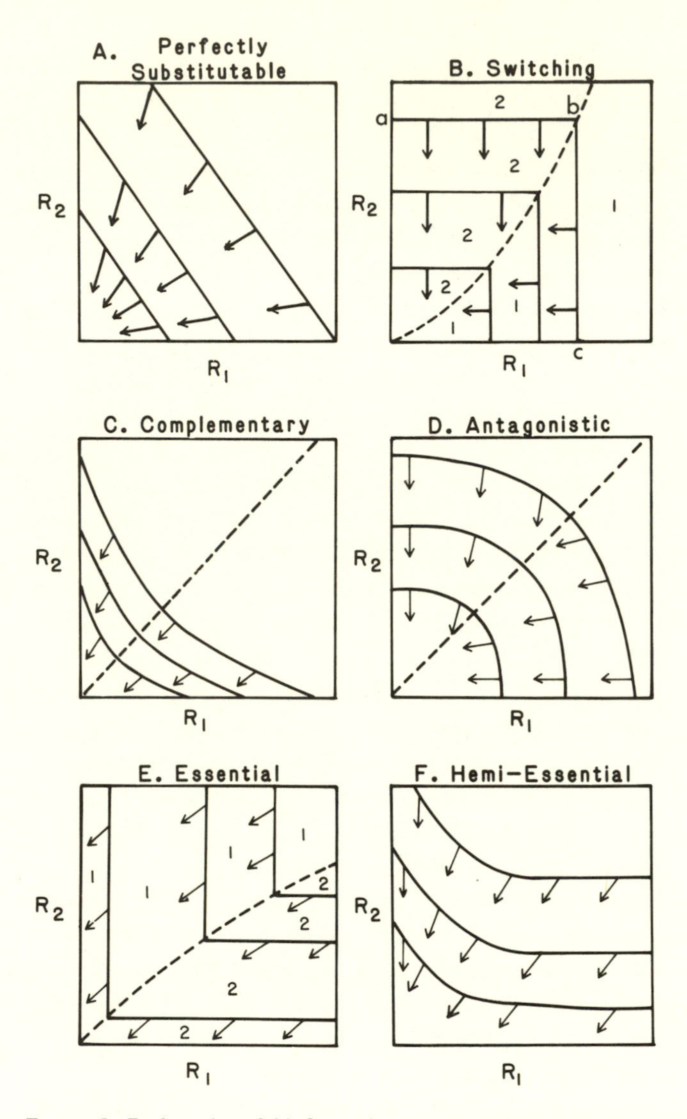

FIGURE 5. Each section of this figure shows resource-dependent growth isoclines and the associated consumption vectors. The growth isoclines show the *environmental availabilities* of R_1 and R_2 for which a population has a given reproductive rate. The vectors represent the amounts of R_1 and R_2 consumed per individual per unit time in a habitat with the resource availabilities at that point on the growth isocline. The slope of each vector is the ratio of the amount of R_2 consumed to the amount of R_1 consumed. These amounts can be determined using a model of optimal foraging, such as those illustrated in Figures 3, 4, 6, 7, and 8.

per unit time. The slope of each of the consumption vectors is the ratio of $R_2 : R_1$ consumed, and the length of each vector is the absolute amount of R_1 and R_2 consumed per unit time. Each vector can be thought of as the sum of two components: one component is the rate of consumption of R_1 and the other is the rate of consumption of R_2. For more details see Figure 20 in Chapter 3.

The linear resource-dependent growth isoclines and associated consumption vectors predicted for nutritionally perfectly substitutable resources with convex consumption constraint curves (Fig. 5A) differ dramatically from those predicted for the cases with either linear or concave consumption constraint curves. As illustrated in Figure 4A and C, these cases lead to switching consumption, with individuals consuming the one resource which leads to the higher reproductive rate. With switching consumption, resource-dependent growth isoclines are no longer straight lines even though resources are nutritionally perfectly substitutable. This can be seen by considering a habitat that gives the curves of Case 1, Figure 4A. In this habitat, only R_2 is consumed. The consumer's reproductive rate is thus determined by R_2, and is independent of R_1. As R_2 is consumed, its availability is decreased relative to R_1. Reproductive rate will continue to be independent of R_1 for increasingly higher relative availabilities of R_1 until R_1 reaches an availability at which an individual has a higher reproductive rate from consumption of R_1 than from consumption of R_2. At this point, the individual will switch to consumption of only R_1, and its reproductive rate will be independent of the availability of R_2. Such a response is represented by switching resource growth isoclines, as illustrated in Figure 5B. To understand the meaning of the switching isocline, consider its two segments. The isocline segment between points *a* and *b* in Figure 5B is the segment for which only R_2 is consumed. For any availability of R_1 along this segment, individuals have a higher reproductive rate from consumption of R_2 than from consumption of R_1. Along the isocline segment from points *b* to *c*, only R_1 is consumed, and growth is independent of R_2 availability. The consumption vectors of Figure 5B show the switch from consumption of only one resource to consumption of only the other resource as availabilities change.

These two examples illustrate the difference between the purely nutritionally based nutrition isoclines and resource-dependent growth isoclines of a species. Although nutrition is one important process influencing the shape of a species' resource-dependent growth isocline, the constraints placed on consumption by habitat structure, foraging tactics, handling time, and resource availability also influence the shape of the resource-dependent growth isocline. Of all the cases considered in this chapter, linear nutrition isoclines are the most sensitive to changes in the shape of the consumption constraint curve. This suggests that the optimal diet of organisms with nutritionally perfectly substitutable resources will be highly dependent on the foraging tactics of individuals and the spatial and temporal structure of the habitat.

Complementary Resources

Two resources, each containing different proportions of two nutritionally essential elements, may lead to a higher growth rate when consumed together than would be predicted for a linearly weighted sum. Such nutritional complementarity has been noted for several foods of herbivores, and has been the subject of much research for both livestock and humans. For instance, humans eating certain types of beans with rice can increase the usable protein content of their food 40% over the sum of the usable protein of beans and rice eaten separately—because beans are rich in lysine, an essential amino acid in low abundance in rice, and rice is rich in sulfur-containing amino acids which are relatively lacking in beans (Lappe, 1971). Such nutritional complementarity occurs only if two foods are consumed within a short time of each other, an important constraint on diet optimization.

For two nutritionally complementary resources, one proportion leads to maximal complementarity at any growth rate. This proportion is shown with a dotted line in Figure 5C. The hypothesized consumption vectors of Figure 5C show a compromise between consumption equal to optimal proportion and consumption proportional to abundance or ease of capture. This is the general trend predicted by the graphical analysis of optimal foraging for nutritionally complementary resources shown in Figure 6. In all cases shown, the analysis predicts a mixed diet. Switching would only be predicted if the concave consumption constraint curve were more concave than the fitness curve (a possibility which is not illustrated). The optimal ratio of R_1 and R_2 in the mixed diet is predicted to depend on the amounts of R_1 and R_2 available, with the proportion of a resource in the diet increasing as its proportionate availability increases. The optimal dietary proportion of a resource changes slowly with changes in the available proportion when there is a convex consumption constraint curve, and changes rapidly with changes in the availability of that resource when there is a concave consumption constraint curve, as inspection of Figure 6 reveals.

Antagonistic Resources

Nutritionally antagonistic resources are substitutable, but have nutrition isoclines which bow away from the origin (Fig. 7). This means that an organism must consume more resources than predicted by a linearly weighted calculation in order to attain a given reproductive rate if two resources are consumed together. Such a phenomenon could be caused by synergistic effects of toxic compounds. For instance, Janzen, Juster, and Bell (1977) demonstrated that non-protein amino acids, such as D, L-pipecolic acid and djenkolic acid, had no significant effect on growth of a bruchid bettle if consumed separately, but had a significant synergistic effect if consumed together. Thus, if one seed contained D, L-pipecolic acid, and

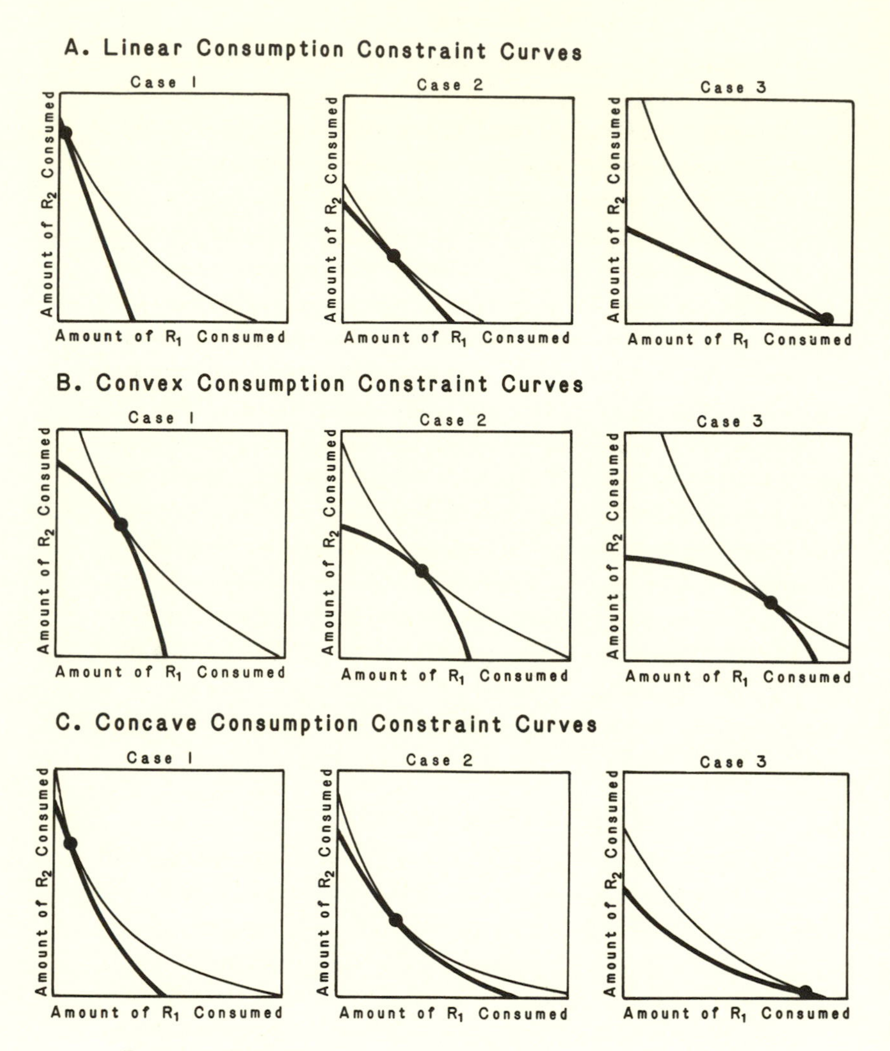

FIGURE 6. Optimal diet for nutritionally complementary resources. Details are similar to Figure 4. Note that the optimal diet is always a mixed diet.

another seed contained djenkolic acid, less growth might be possible from consumption of the two seeds together than from a comparable consumption of either the one seed or the other. The known synergisms of various drugs used in human medicine (many being secondary plant compounds) also suggest that nutritional antagonisms may occur. However, spatial heterogeneity may be a more likely cause of resource antagonism.

If two resources are antagonistic, a diet biased toward mainly one or the other, i.e., specialization, would seem beneficial. Intermediate combinations would lead to greater antagonistic effects, requiring a greater absolute food intake to achieve the same per capita reproductive rate. Optimal diet for antagonistic resources is explored in Figure 7. For the linear, convex, and concave consumption constraint curves shown, dietary switching is predicted. This would lead to resource-dependent growth isoclines and consumption vectors as shown in Figure 5B. If the amount of nutritional antagonism were small relative to the convexity of the consumption constraint curve (a case not illustrated in Figure 7), a mixed diet would be possible. This could lead to the growth isoclines and consumption vectors illustrated in Figure 5D.

Essential Resources

In order to live, all organisms require an energy source and various forms of the elements N, P, C, O, H, S, Fe, K, Ca, etc. Autotrophs most often obtain these elements separately from their environment. Heterotrophs, being consumers of autotrophs, obtain these in more complex forms, with several basic nutritional elements in the same food item. For this reason, the majority of resources of autotrophs are essential, whereas those of heterotrophs are likely to be substitutable or hemiessential.

Numerous studies have demonstrated that inorganic plant nutrients are essential (see any introductory plant physiology text). A few recent experiments have indicated that plant

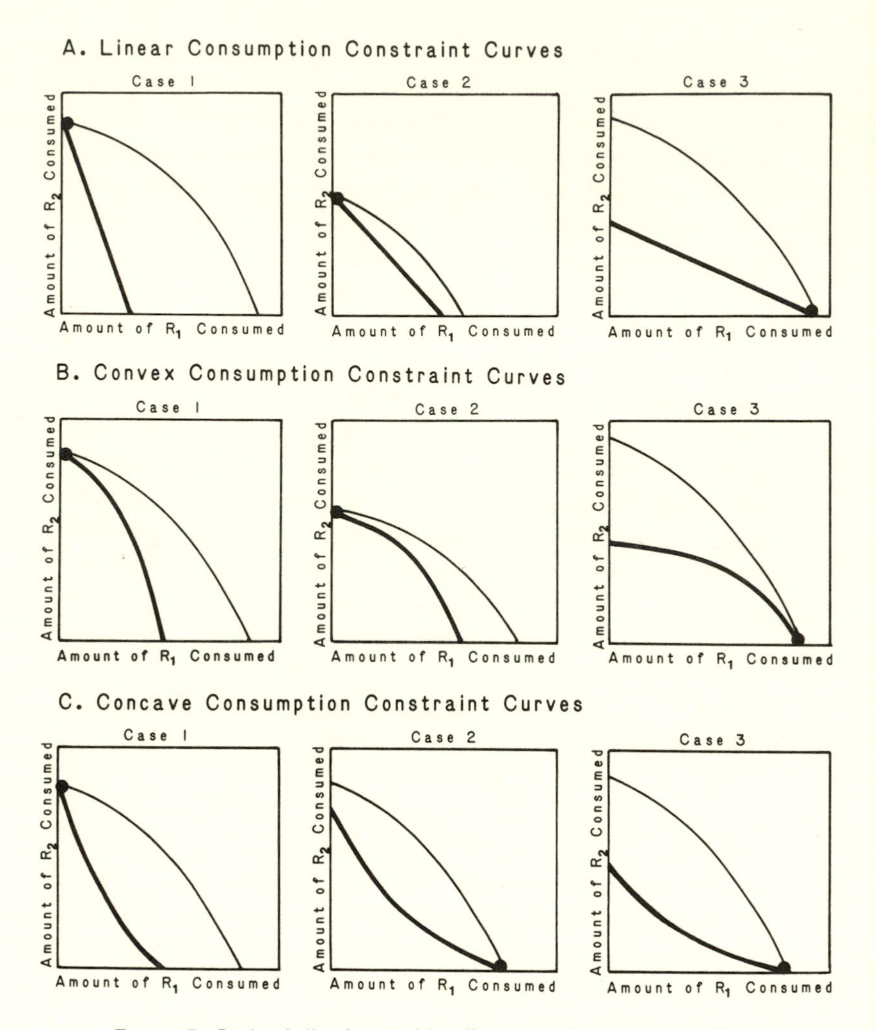

FIGURE 7. Optimal diet for nutritionally antagonistic resources. Note that switching consumption is predicted for all cases.

nutrients may be non-interactive essential resources, i.e., give isoclines with a right-angle corner. Droop (1974) grew a marine alga with vitamin B-12 and phosphate as potentially limiting nutrients, and found that his results fit the non-interactive essential model. Rhee (1978) similarly demonstrated that phosphate and nitrate were non-interactive essential resources for a freshwater alga. Because the available experimental evidence indicates that plant nutrients are non-interactive essential resources, I will call non-interactive essential resources simply "essential resources," but maintain the longer description for interactive essential resources.

Not all pairs of plant nutrients are essential with respect to each other. Consider N_2, $NO_2^{=}$, NO_3^{-}, NH_4^{+}. Each of these four forms of nitrogen can be used by some plants as the sole source of nitrogen, and may be substitutable relative to the others. N_2 is a nitrogen source only for species capable of nitrogen fixation—mainly legumes and blue-green algae. Because of the high energetic cost of reducing atmospheric N_2 and because of the need to induce specialized systems to do this, it might be that N_2 will be found to be antagonistic or even switching relative to the reduced forms of nitrogen which may be used by nitrogen-fixing organisms. Thus, not all pairs of plant nutrients are likely to be essential resources. However, all usable forms of nitrogen-containing compounds are likely to be essential relative to all usable forms of phosphorus-containing compounds. Such considerations allow the pairwise definitions offered here to be extended to groups of resources.

For a pair of essential resources, the growth rate of a species will be determined by either one or the other resource, whichever is "more limiting." The values of R_1 and R_2 for which a species is equally limited by both resources are defined by a curve from the origin through the corners of the resource growth isoclines (the broken curve of Figure 5E). Above this curve, in the region labeled 1, the species is limited by R_1. Below it, the species is limited by R_2. The species is equally limited by both resources for points along this curve. The slope of this curve at any given growth rate is the optimum ratio of resources which is required for growth at that rate. The curve shown with a broken line in Figure 5E gives the optimal ratio of R_1 and R_2 at a given reproductive rate. This curve may bend, as shown, or be a straight line.

At a particular growth rate, it would seem that the consumption rate of R_1 and R_2 would be proportional to the tangent to this curve of optimum proportion at that growth rate. This is to say, resources would be consumed in the proportion they are required for long-term, balanced growth. If one resource were consumed in excess of this proportion, there would be no gain in current reproductive rate (unless excess consumption were detrimental to a competitor or resource availability fluctuated), and there would be a loss if such excess consumption had a reproductive cost. The cost of excess consumption of one resource would be reflected in the position of the growth isocline of an individual with such a trait. The greater the reproductive cost of excess consumption of a non-limiting resource, the further from the origin would be the growth isocline of that individual. As is shown in Chapter 4, such an individual would be at a competitive disadvantage compared to other individuals within the species which did not consume excess quantities of a non-limiting resource. Instantaneous rates of resource consumption consistent with this hypothesis are shown in Figure 5E using vectors. Each of the vectors shows the amounts of R_1 and R_2 consumed per unit time. The slope of each vector is the optimal proportion of the resources, which may also be expressed as the optimum ratio, R_2:R_1.

This conclusion is identical to that provided by the graphical model of optimal diet. For linear, convex, and concave consumption constraint curves, two nutritionally essential resources should be consumed in the proportion in which growth is equally limited by the two resources (Figure 8). Such consumption leads to the isoclines and vectors of Figure 5E. Consumption

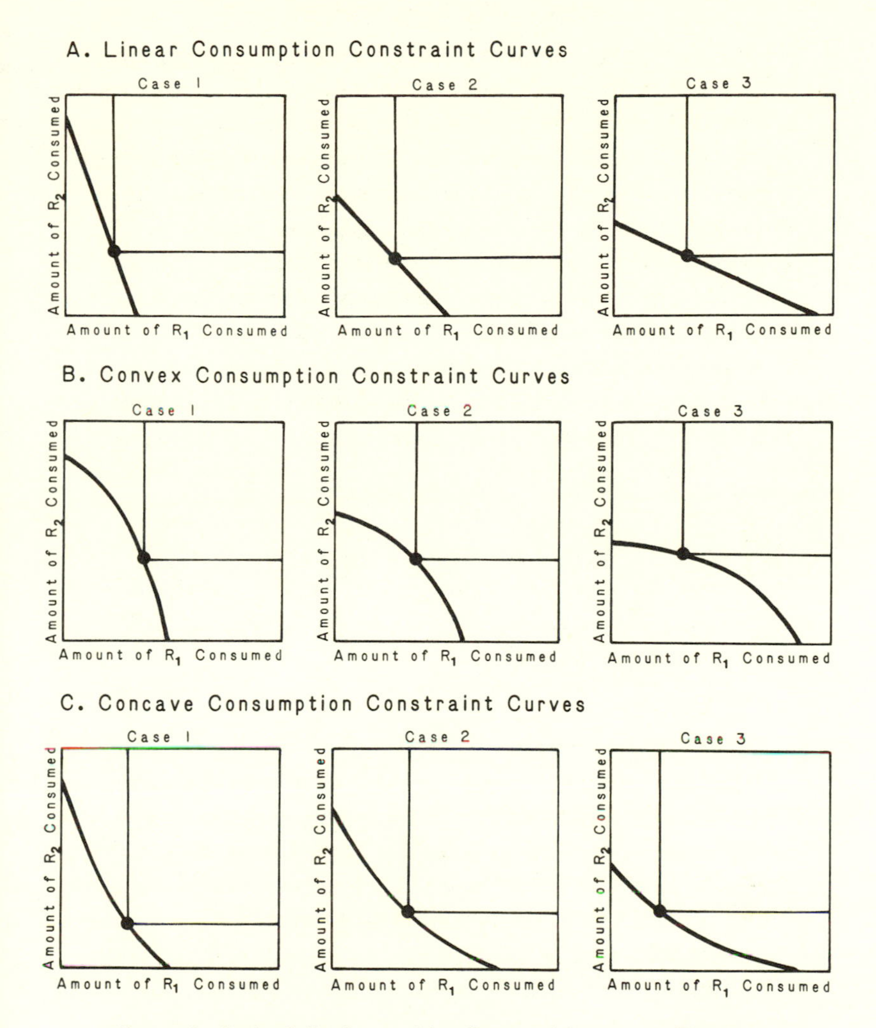

FIGURE 8. Optimal diet for nutritionally essential resources. The predicted optimal diet is consumption of R_1 and R_2 in the ratio at which the resources equally limit growth, i.e., in the ratio which occurs at the kink point of the nutrition isocline.

vectors comparable to those of Figure 5E will be used throughout the rest of this book for all cases of consumption of essential resources because the optimal diet of essential resources is predicted to be independent of the shape of consumption constraint curves. For all cases, essential resources should be consumed in the proportion at which growth is equally limited by both resources. However, "bet hedging" and environmental variance (Levins, 1979, Armstrong and McGehee, 1980; Chapter 9) could favor hoarding.

As noted by Leon and Tumpson (1975) and used by Tilman (1977), essential resources may be easily modeled. The equation governing population growth for species i is

$$dN_i/N_i\,dt = \underset{j=1,k}{\mathrm{MIN}}\,(f_i(R_j) - m_i). \tag{3}$$

The forms of $f_i(R_j)$ define the shape of the optimal proportion curve and the placement of the growth isoclines. As shown in Figure 1, $f_i(R_j)$ may have the shape of the Monod (1950) function or the Type I, II, and III curves of Holling (1959).

Hemi-essential Resources

A pair of resources may be nutritionally hemi-essential if one is nutritionally complete and the other lacks some nutritional element or elements available in relative excess in the first resource. For a herbivore, some fruits may be hemi-essential relative to foods such as seeds. Many fruits are rich in carbohydrates, fats, or oils, but low in protein, and may lack one or more nutritionally required amino acids. No reproduction would be possible on a long-term diet of only such fruit. However, the caloric content of the fruit would allow it to supplement a diet of other foods which were relatively rich in protein, as are many seeds and animals.

Figure 9 illustrates the cases of optimal diet for hemi-essential resources. In all cases, the graphical theory predicts a mixed diet, with the optimal proportions of R_1 and R_2 in the diet

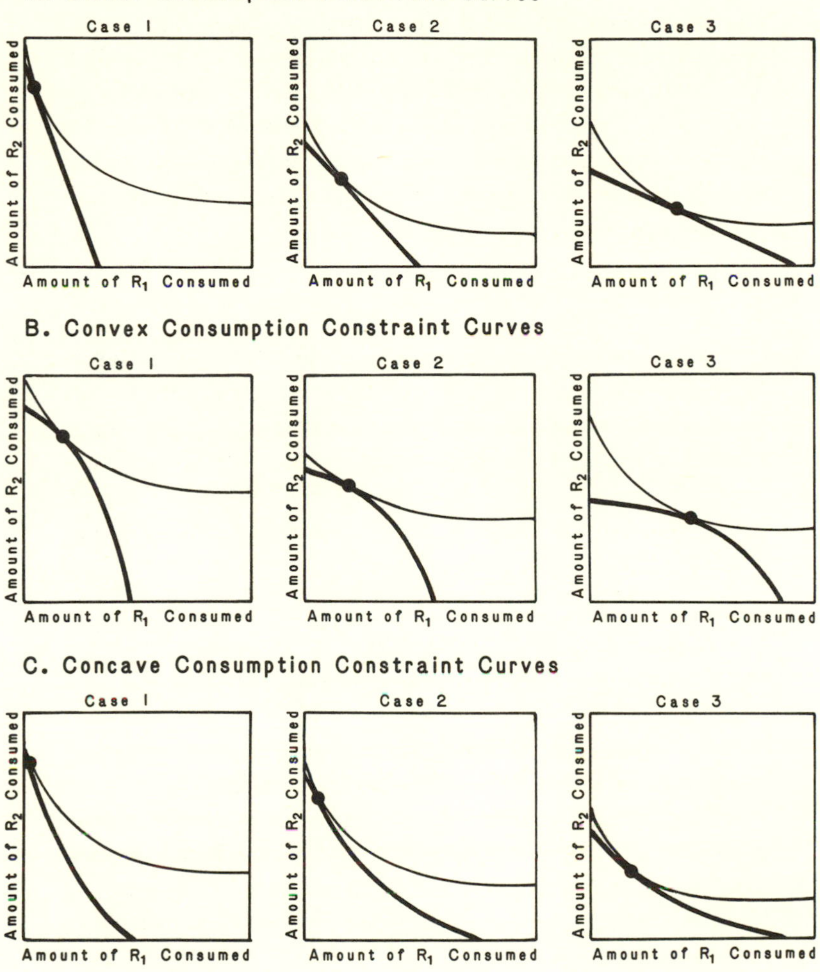

FIGURE 9. Optimal diet for nutritionally hemi-essential resources. The predicted optimal diet is always a mixed diet. with the dietary proportion of a resource increasing as the availability of the item increases.

changing with their relative availabilities. For linear, convex, and concave curves (Fig. 9), increasing availability of R_1 (the non-essential resource) leads to an increase of R_1 in the diet. The most rapid increase of dietary R_1 occurs with concave consumption constraint curves, and the slowest change in dietary proportion occurs with convex curves. These trends are illustrated with the resource-dependent growth isoclines and consumption vectors of Figure 5F.

Switching Resources

The resource-dependent growth isoclines representative of switching resources, as already discussed, result from the interactions between consumption and nutritional constraints. Given the proper consumption constraints, nutritionally substitutable resources may be consumed in a switching manner, giving switching growth isoclines. Switching is easily modeled:

$$dN_i/N_i\,dt = \max_{j=1,k} \left(f_i(R_j) - m_i\right). \tag{4}$$

For switching resources, growth rate is determined solely by the availability of the one resource which leads to the largest reproductive rate.

The broken line in Figure 5B indicates the proportions of R_1 and R_2 at which two switching resources lead to the same reproductive rate. Above this line, in the area labeled 2, only R_2 would be consumed, because R_2 leads to the higher growth rate. In region 1, only R_1 would be consumed, because R_1 leads to the higher growth rate. Such consumption vectors are shown in Figure 5B.

Even in the absence of nutritional antagonism, selection for foraging ability may lead to near perfect switching, as already discussed. The ecological factors leading to such switching may become clearer when a specific situation is discussed in more detail. Consider an individual foraging for two resources which are nutritionally perfectly substitutable, but which exist in a spatially heterogeneous environment. If each resource occurs in a pure patch, a consumer will be able to forage for only one of the resources within any small time interval. An efficient forager might choose to forage in the patch with resource availability that leads to the higher net reproductive rate. When patches of this resource become depleted, it would switch to the patches of the other resource. Switching can also occur when resources are distributed uniformly in a habitat, as illustrated by Figure 4A. Foraging behavior approaching this and relevant theory has been reported by Murdoch (1969, 1971), Murdoch and Marks (1973), Murdoch *et al.* (1975), and others.

Another factor leading to switching can be found in the different behavioral or physiological traits which are required to forage for different resources. Heinrich (1976a, 1976b, 1979) has documented such switching behavior (majoring) for individual bumblebees. The phenomenon of sequential utilization of sugars (diauxy) in bacteria is also a form of switching, caused by the need for different enzymes to use each sugar.

On nutritional grounds, it might seem that most resources of motile heterotrophs should be approximately nutritionally perfectly substitutable, with some being either slightly complementary or slightly antagonistic. This nutritional argument was given by MacArthur (1972) and has been used by others to justify using the linear growth isoclines of perfectly substitutable resources as the best initial approximation for the resource class of motile animal species. However, the graphical exploration of optimal diet in Figure 4 suggests that this is not the best initial approximation. For two of the three possible cases, nutritionally perfectly substitutable resources should give switching resource-dependent isoclines. In the other case, linear isoclines are expected. The less convex the consumption constraint curve and the more convex the nutrition isocline, the greater the tendency for nutritionally substitutable resources to give switching growth isoclines. Concave consumption constraint curves are caused by spatially and temporally patchy distributions of resources and by a tendency for foragers to use a search image. The prevalence of these factors, at least for consumers of immobile prey, suggests that the majority of resources of motile animals will be responded to in a switching manner. This possibility, and its implications for the structure of animal communities, will be considered in Chapter 9. The analyses of Figures 4, 6, and 7 suggest that animals which do not respond to resources in a switching manner would most likely be those with nutritionally complementary resources. Thus, the second most common type of animal resources should be complementary resources.

SUMMARY

A resource is defined as a consumable factor for which increases in its availability lead to increased per capita reproductive rates through at least some range of its availability. Resource-dependent growth isoclines (Fig. 2) are used to classify pairs of resources as being substitutable, essential, or hemi-essential. Substitutable resources may be perfectly substitutable, in which case they would give linear growth isoclines; or complementary, in which case the growth isoclines would bow in toward the origin; or antagonistic, in which case the isoclines would bow away from the origin; or switching, in which case the isoclines would form a right-angle corner bowed away from the origin. A simple graphical model of optimal foraging indicates that the shape of the resource-dependent growth isocline reflects both nutritional and ecological processes. The analysis suggests that resources which are nutritionally perfectly substitutable should often be consumed in a way which gives switching growth isoclines, and that essential resources should be consumed in the proportion for which the population would be equally limited by both resources, called the optimal ratio.

CHAPTER THREE

Competition for a Single Resource

When several species are limited by the same resource, the dynamics and long-term outcome of competition can, in theory, be predicted using information on the resource requirements of the species and the supply rate of the resource. The long-term outcome of competition may be either a true equilibrium, in which the rates of change of all species and the resource are zero, or it may be an oscillatory solution in which resource level and population densities fluctuate, with the population density of one or several species being bounded away from zero. This chapter is concerned with the mechanism of competition when a true equilibrium is reached, i.e., when $dN_i/dt = dR/dt = 0$. Cases which lead to sustained resource oscillations are intriguing, and are discussed in Chapter 9. As discussed in Chapter 9, the long-term outcome of competition for a single fluctuating resource can be interpreted using the graphical, equilibrium theory of competition for two resources developed in Chapter 4. This occurs because resource fluctuations may be responded to as if they were a resource, or as if they were a limiting factor (Levins, 1979; Armstrong and McGehee, 1980).

The mechanism of competitive displacement when several species compete for the same resource is assumed to be through depression of resource availability. This is best illustrated by first considering the interaction of a single species with a single limiting resource. Consider a species with a resource-dependent population growth curve as shown in Figure 10, which experiences the per capita mortality rate (m) shown. For this species

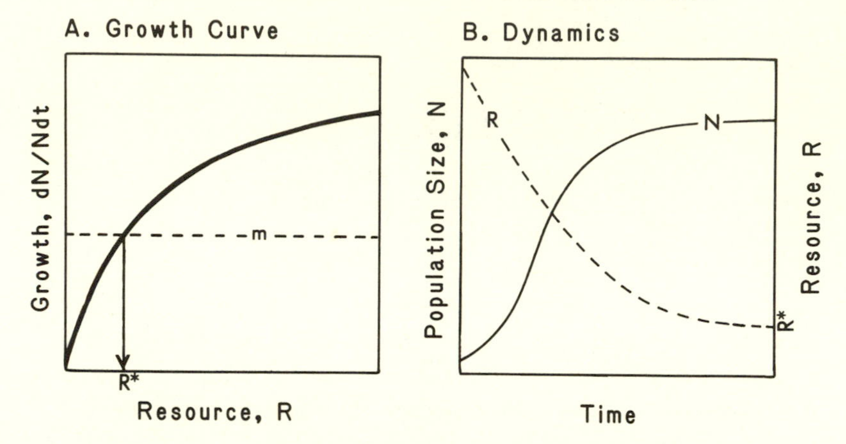

FIGURE 10. A. The solid curve shows the dependence of per capita reproductive rate on resource availability. The broken curve labeled *m* is the mortality rate experienced by this species. The resource availability at which these two curves cross is R^*, which is the resource availability at which reproductive rate equals mortality rate. R^* thus represents the amount of resource that this species requires in order to maintain a stable equilibrium population.

B. The solid curve shows the population dynamics and the broken curve the resource dynamics which could result from growth of the species of part A of this figure in a habitat with a mortality rate of *m*. Note that the population density approaches an equilibrium value as the resource level approaches R^*.

to maintain a stable, equilibrium population at this mortality rate, it must have sufficient resource to grow at a rate equal to this mortality rate. The amount of resource needed to do this, termed R^*, is easily determined graphically (see Fig. 10A). (Throughout this book, a superscript asterisk will be used to represent equilibrium quantitites of a variable.) As illustrated, the growth curve and mortality rate of a species determine its resource requirement at equilibrium, R^*. If this species were to invade a habitat in which it experienced a mortality rate of *m*, and resources were initially more abundant than R^*, its population density would increase. As the population grew, the resource level would decline. The level would continue to decline, and population density would continue to increase,

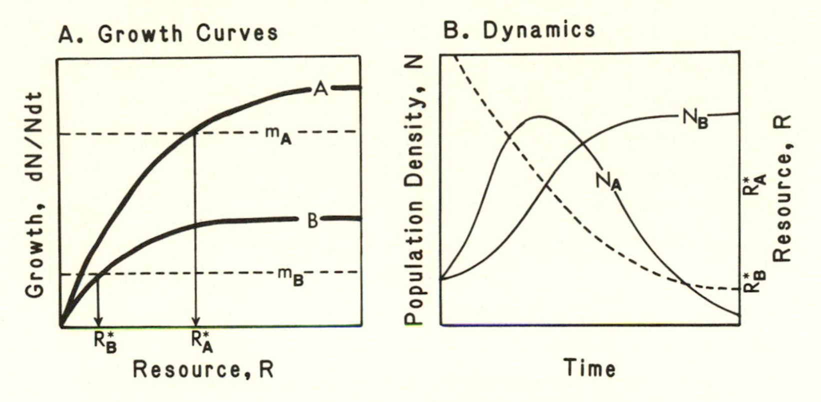

FIGURE 11. A. The two solid curves labeled A and B show the resource-dependent growth curves for species A and B. The broken lines labeled m_A and m_B show the mortality rates for species A and B. The equilibrium resource requirements of these species calculated using these two curves are labeled R^*_A and R^*_B.

B. The solid curves show the population dynamics for species A and B, and the broken line shows the nutrient dynamics. Species B approaches an equilibrium population density, while species A is competitively displaced after an initial period of growth. This occurs because $R^*_B < R^*_A$. Species B is able to reduce the resource to R^*_B, at which value there is insufficient resource for the survival of species A.

until resource levels reached R^*. At R^*, the growth rate of the population would exactly balance its mortality rate. If the supply rate of the resource exactly balanced the consumption rate, this would be an equilibrium. Such a case is illustrated in Figure 10B. Whether such an equilibrium would be reached, either asymptotically or with damped oscillations, and the stability of the equilibrium would depend on initial conditions and on the functions describing the growth response of the species and the resource supply process. When a stable equilibrium is reached, the species has reduced resource levels down to its R^*. R^* is thus the amount of resource which the species must have in order to maintain a stable, equilibrium population in a habitat.

Figure 11 graphically explores a case of interspecific competition for a limiting resource. As shown in Figure 11A, species A

has a higher R^* than does species B. If resource levels were initially greater than R_A^*, both species would increase in population density (Fig. 11B). As population sizes increased, resource levels would be decreased. The population size of species A would stop increasing when the resource level was decreased to R_A^*, but the population size of species B could continue increasing until resource levels were reduced to R_B^*. At this level, species A would have insufficient resources to maintain a stable population. It would be competitively displaced by species B, as shown.

These graphical treatments suggest that, when several species compete for the same limiting resource, the one species with the lowest equilibrium resource requirement (R^*) for the limiting resource should competitively displace all other species at equilibrium (see also O'Brien, 1974; Hsu *et al.*, 1977). Amstrong and McGehee (1980) present an elegant mathematical proof of this assertion.

As an explicit treatment of competition among several species for one resource, consider a model based on the Monod (1950) equation. The Monod equation is a good approximation to the growth function of many species. The Monod function is illustrated in Figure 12. The maximal growth rate ($dN/N\,dt$) asymptotically reached by a species is termed r, and the resource availability at which growth reaches half of the maximal growth rate is termed k, the half saturation constant. If several species respond to resources as defined by the Monod model, the equations of competition among n species for one resource are

$$dN_i/N_i\,dt = r_i R/(R + k_i) - m_i,$$

$$dR/dt = a(S - R) - \sum_{i=1}^{n} (dN_i/dt + m_i N_i)/Y_i, \qquad (5)$$

where the subscript i refers to species i, m_i is the mortality rate of species i, Y_i is the number of individuals of species i produced per unit resource, S is the amount of resource being supplied

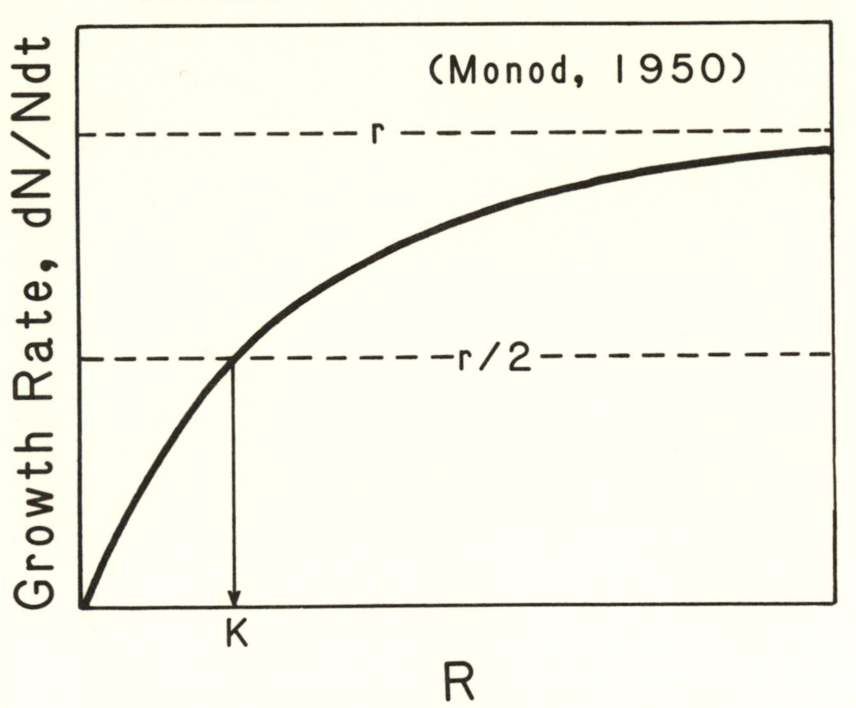

FIGURE 12. The solid curve shows the Monod growth function. The per capita reproductive rate is zero when no resource is available, and increases in a saturating manner with resource availability. The asymptotically approached maximal growth rate is r. In order to grow at half of its maximal rate, the species requires an amount of resource called the half saturation constant, represented by k.

to the system, and a is the rate constant for resource supply. As given, these equations assume the resource supply process of a "chemostat." At equilibrium ($dN_i/N_i\,dt = 0$ and $dR/dt = 0$), these equations give

$$R_i^* = k_i m_i/(r_i - m_i), \qquad (6)$$

with one value of R_i^* for each species. Theory predicts that the species with the lowest R^* will competitively displace all other

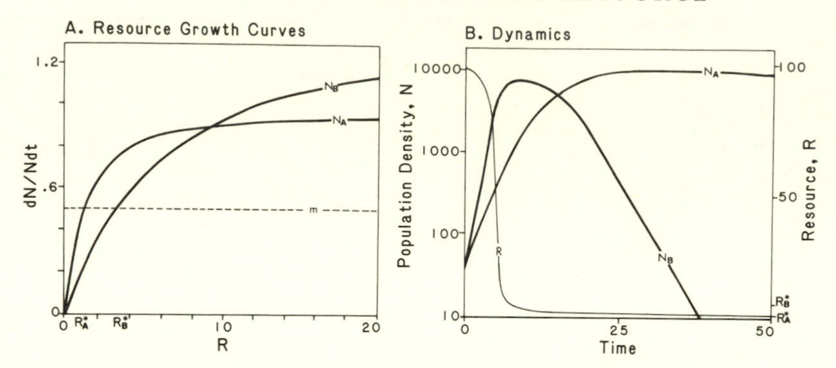

FIGURE 13. This figure illustrates an explicit solution of the Monod model of competition for a single resource. Part A shows the growth curves and mortality rate used, and part B gives the population and resource dynamics predicted by the model. Because R_A^* is less than R_B^*, species *A* competitively displaces species *B* by reducing the resource to a level at which species *B* cannot maintain a stable population.

species at equilibrium, independent of initial population densities. Two species may coexist at equilibrium when limited by the same resource only if they have identical R^*. The graphical example presented in Figure 11 shows the mechanism of competitive displacement, but does not prove this assertion. A thorough mathematical proof that the species with the lowest resource requirement as measured by R^* will be the superior competitor at equilibrium has been given by Hsu, Hubbell, and Waltman (1977), and will not be repeated here. However, the basis of the proof is easily seen from examining Figure 13 and Equations 5 and 6. The species of Figure 13 follow Eq. 5, and their parameters are shown in the figure. The initial value for R is 100. At this resource availability, examination of Eq. 5 shows that the population density of each species will be increasing. Equation 5 also shows that as population densities increase, R will eventually decrease. When R reaches a value of 3, $dN_A/N_A\,dt = +0.75$, while $dN_B/N_B\,dt = 0$, because $R_B^* = 3$. The growth of species A will further reduce R. The population density of species A will continue to increase and the resource availability will continue to decrease until R is reduced to R_A^*. At $R = R_A^* = 1$, $dN_A/N_A\,dt = 0$, and $dN_B/N_B\,dt = -0.286$. The equilibrium density of species A will be $N_A^* = aY_A(S - R_A^*)/m_A = 9900$, at which population density $dR/dt = 0$. The species with the lower R^* competitively displaces the other species because it reduces the level of the limiting resource below that required for the other species to maintain a stable population.

When numerous species compete for one limiting resource, their R^* may be used to rank their predicted competitive ability. To do this, renumber the competing species so that

$$R_1^* < R_2^* < R_3^* < R_4^* < \cdots < R_n^*. \tag{7}$$

When the species are so ranked, species 1 is predicted to competitively displace all other species at equilibrium, species 2 is predicted to competitively displace all species except species 1 at equilibrium, etc. According to theory, the equilibrium outcome of competition should be the same independent of the initial population densities of the competing species. The equilibrium density of the competitively dominant species will be

$$N_1^* = aY_1(S - R_1^*)/m_1, \tag{8}$$

and the equilibrium density of all other species will be $N_i^* = 0$, for $i \neq 1$.

EXAMPLES OF COMPETITION FOR A SINGLE RESOURCE

Tilman, Mattson, and Langer (1981) tested the predictive ability of the Monod model of competition for a single limiting resource using two species of freshwater algae isolated from Lake Michigan. The two species were both diatoms, a group of aquatic microscopic plants which require silicate (SiO_2) for a cell-wall structure called a frustule. Diatoms are often the major producers of mid-latitude, moderately productive lakes. One

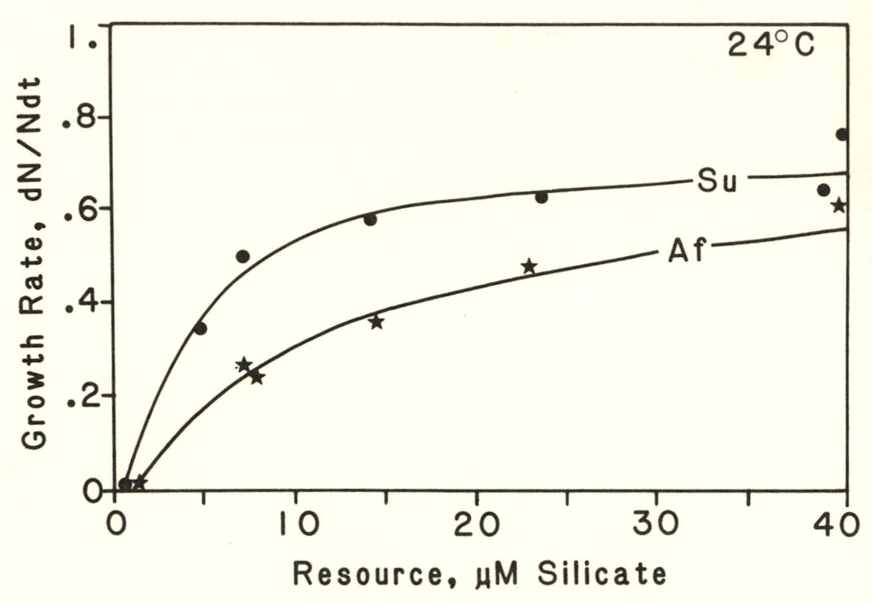

FIGURE 14. These two curves show the dependence of the per capita reproductive rate of *Synedra ulna* (Su) and *Asterionella formosa* (Af) on silicate concentration, as determined using short-term batch culture experiments. Data are from Tilman, Mattson, and Langer, 1981.

of the species studied (*Asterionella formosa*) is one of the more dominant algae of Lake Michigan, whereas the other species (*Synedra ulna*) is limited to nearshore waters. Studies on natural algal communities have indicated that the growth of many diatom species is determined by silicate concentrations in the lake (Kilham, 1971; Schelske, 1975).

Figure 14 shows experimentally determined silicate-limited growth curves for these two species at 24°C. The information from such short-term batch cultures can potentially predict the outcome of long-term competition experiments. The long-term experiments were performed in semicontinuous-flow chemostats (Tilman, 1977). When grown together in a chemostat in which new medium is added at a constant rate and old medium and algal cells are removed at the same rate, each species experiences a mortality rate equal to the dilution rate. For these experiments this was 0.11 day $^{-1}$. This mortality rate gives an R^* for *Synedra* of 1.0 μM SiO_2 and an R^* for *Asterionella* of 2.8 μM SiO_2. The population dynamics of *Asterionella* and *Synedra* when grown alone at a dilution (mortality) rate of 0.11 day^{-1} are shown in Figure 15A and B. *Asterionella* smoothly approached an asymptotic population maximum and reduced the extracellular silicate concentration at equilibrium to about 1.0 μM. *Synedra* also reached an equilibrium and reduced extracellular silicate concentrations to 0.4 μM. Although the equilibrium levels differed from the R^* predicted using the short-term growth experiment's parameters, *Synedra* did reduce silicate concentrations to a lower point than did *Asterionella*. Both the actually observed values for R^* (0.4 μM SiO_2 for *Synedra* and 1.0 μM SiO_2 for *Asterionella*) and the values predicted from the short-term experiments of Figure 14 (1.0 μM SiO_2 for *Synedra* and 2.8 μM SiO_2 for *Asterionella*) predict that *Synedra* should competitively displace *Asterionella* when both compete for silicate at a mortality rate of 0.11 day^{-1} at 24°C.

When these two species did compete (Fig. 15C, D, and E), *Synedra* displaced *Asterionella* independent of the initial density of either species, as predicted by theory. The dynamics of the winning species, *Synedra*, closely followed those predicted by the Monod model, and those of *Asterionella* were fairly similar to predictions. It should be noted that, in all cases, extracellular silicate concentrations were reduced to about 0.4 μM, sufficient for *Synedra* to maintain an equilibrium population as demonstrated by the single species experiment of Figure 15A, but insufficient for *Asterionella*. Also note that *Synedra* displaced *Asterionella* when both species had the same initial density (Fig. 15C), when *Asterionella* was initially more abundant than *Synedra* (Fig. 15D), and when *Synedra* was initially more abundant than *Asterionella* (Fig. 15E), as predicted by the theory of competition for a single resource. This example supports the

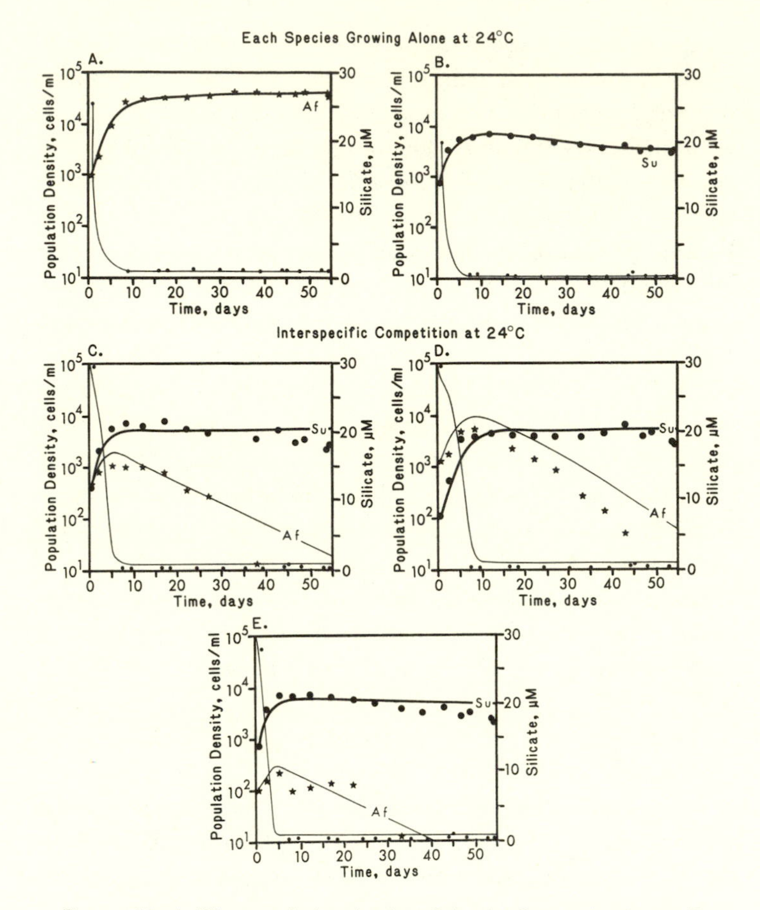

FIGURE 15. A. The population density of *Asterionella* growing by itself in the semi-continuous culture flask (stars)—and the concentrations of silicate, the limiting nutrient (dots), in this flask—are graphed against time.

B. Population dynamics of *Synedra* growing by itself (large dots) and concurrent dynamics of silicate concentrations (small dots) under the same conditions as for *Asterionella* in part A.

C, D, and E. Competition between *Asterionella* (stars) and *Synedra* (dots) under conditions of limiting silicate for three different starting densities of the two species. In all cases, *Synedra* competitively displaces *Asterionella*, as predicted using the information of Figure 14. (Data from Tilman, Mattson, and Langer, 1981.)

ability of a simple model of resource competition, the Monod model, to predict the long-term equilibrium outcome and the population dynamics of competition.

Similar short-term and long-term growth experiments were performed with *Asterionella* and *Synedra* at other temperatures ranging from 4 to 24°C. The results of the short-term experiments were used to predict the temperature dependence of R^* for these two species along this temperature gradient, for a mortality rate of $m = 0.11\ \text{day}^{-1}$. The predicted R^* are shown in Figure 16A for *Synedra* and in Figure 16B for *Asterionella*. The solid line in each of these figures is the resource-temperature isocline along which the reproductive rate of each species would be $0.11\ \text{day}^{-1}$. This isocline defines the amount of silicate that each species requires in order to maintain a stable population density at a given temperature, when experiencing a mortality rate of $0.11\ \text{day}^{-1}$. For silicate concentrations and temperature in the shaded region of each figure, population density should increase. Population density should decrease for points in the unshaded region.

Figure 16C shows how these isoclines may be used to predict the equilibrium outcome of silicate competition between these two species at various temperatures. The isoclines cross at approximately 20°C. Above this temperature, *Synedra* requires less silicate to maintain itself in such a habitat, and below this temperature *Asterionella* requires less silicate. Thus, *Asterionella* should be the superior competitor below 20°C, and *Synedra* the superior competitor above 20°C. The two species should be able to coexist at equilibrium when limited by this one resource only at the temperature at which they have exactly identical resource requirements, approximately 20°C. The results of competition experiments performed at 8, 13, 20, and 24°C are generally consistent with these predictions. *Asterionella* did displace *Synedra* for all starting conditions at 8 and 13°C. *Synedra* displaced *Asterionella* at 20 and 24°C (Tilman *et al.*, 1981). Although the short-term growth experiments (as in Fig. 14) suggested that

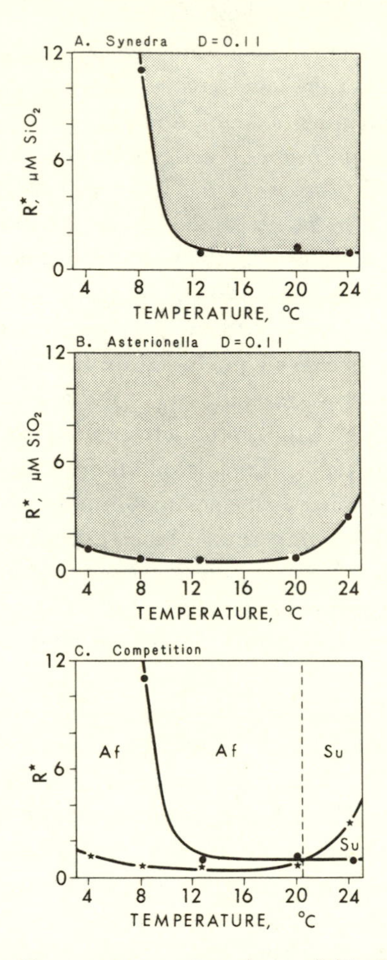

FIGURE 16. A. The temperature dependence of R^* of *Synedra* for a mortality rate of 0.11 day^{-1}. The temperature-silicate growth isocline shows the conditions needed for *Synedra* to maintain a stable population. If environmental conditions have temperatures and silicate concentrations in the shaded region, the population density of *Synedra* will increase. In the unshaded region, *Synedra*'s density will decrease. Equilibrium only occurs for a point on the isocline.

B. The same relationship is shown for *Asterionella*.

C. By superimposing these two growth isoclines, the outcome of silicate competition between these two species at various temperatures can be predicted. The species with the lower requirement (R^*) for the limiting resource (silicate) at a given temperature should competitively displace the other species. Thus, *Asterionella* should be dominant below ca. 20°C, and *Synedra* should be dominant above ca. 20°C. Results of numerous competition experiments at various temperatures are generally consistent with these predictions (Tilman, Mattson, and Langer, 1981).

Asterionella and *Synedra* had almost identical resource requirements at 20°C, the long-term cultures (as in Fig. 14A and B) suggested that *Synedra* had a lower resource requirement at this temperature, and it was dominant.

These results illustrate one way in which knowledge of the resource requirements of species can be used to predict how a resource and a physical factor will interact to determine the patterns of dominance of competing species. Figure 16 is very similar to the usual niche diagrams. The boundary of the niche is determined by the temperature dependence of the resource requirement of each species. The curves of Figure 16A and B show the fundamental niche of each species. Resource competition theory predicts the outcome of competition in the various regions of niche overlap, and gives the realized niches illustrated in Figure 16C.

Braakhekke (1980) provides an interesting example of competition between two terrestrial flowering plants (*Plantago lanceolata* and *Chrysanthemum leucanthemum*) for limiting calcium. Because of the long life span of these plants, the outcome of resource competition was inferred using the replacement principle of deWit (1960). The roots of the two species were suspended in large flasks of deionized water to which all nutrients except calcium were added in excess. Because there was no nutrient replacement, the species competed for the limiting calcium. Using Braakhekke's data on the monocultures of each species, it can be calculated that *Plantago* was capable of reducing Ca down to a level of about 0.4 meq/L but that *Chrysanthemum* could only reduce Ca to about 0.8 meq/L. Resource competition theory would predict that *Plantago* would be the superior calcium competitor because of its ability to reduce Ca to a level below that required for the survival of *Chrysanthemum*. Braakhekke's competition experiments showed that *Plantago* was competitively dominant over *Chrysanthemum*.

Another example of competition for a single limiting resource comes from the work of Hansen and Hubbell (1980). Using various strains of the bacteria *Escherichia coli* and *Pseudomonas aeruginosa*, they first determined the R^* of each strain for the limiting resource, tryptophan, using short-term batch cultures. These R^* were used to predict the outcome of competition for this limiting resource in chemostat culture. Their various experiments (Fig. 17) showed that the superior competitor was not always the species with the higher maximal growth rate (r), nor was it always the species with the lower half saturation constant (k), but that the species with the lower R^* for the limiting resource, tryptophan, always competitively displaced the other species. Hansen and Hubbell (1980) were also able to test another hypothesis that two species limited by the same resource could coexist at equilibrium if they had identical R^*. By varying the concentration of nalidixic acid, which inhibited the growth of one strain but not the other, they were able to demonstrate long-term coexistence only at that concentration of nalidixic acid for which the predicted R^* of the two species for the limiting resource, tryptophan, was identical (Fig. 17C and D).

In another test of the predictive ability of the R^* criterion, Tilman (1977) showed that, for two species of freshwater diatoms competing for either limiting silicate or limiting phosphate, the species with the lower requirement for the limiting resource, as measured by R^*, was always the superior competitor. The freshwater diatom *Asterionella formosa* was shown from single-species growth experiments to have a significantly lower R^* for phosphate compared to *Cyclotella meneghiniana*, and *Asterionella* was the superior competitor for cases of phosphate competition. *Cyclotella* was shown to have a significantly lower R^* for silicate, and it was the superior competitor under silicate limitation. Other aspects of these experiments will be discussed later.

Another set of experiments on Lake Michigan diatoms (Tilman, 1981) showed that two species, *Asterionella formosa* and *Fragilaria crotonensis*, had identical R^* for both silicate and

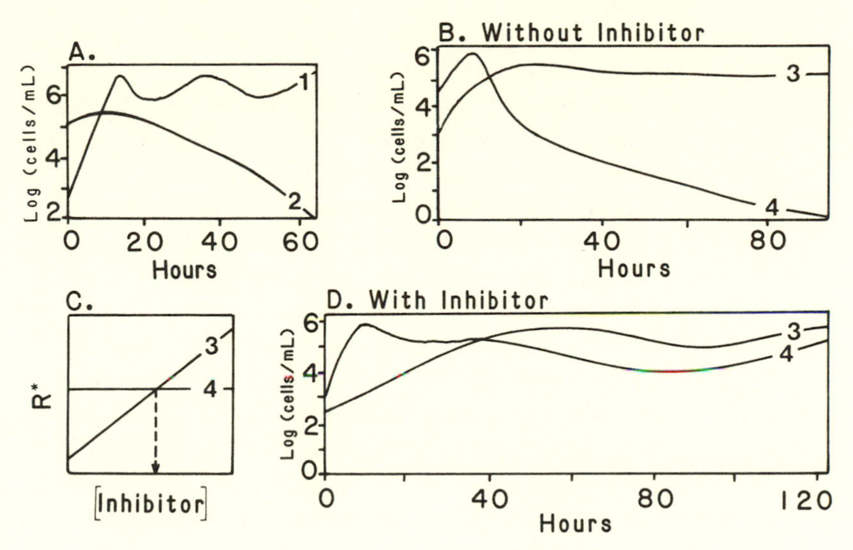

FIGURE 17. Competition for a single limiting resource between various bacterial clones (from Hansen and Hubbell, 1980). Population density is graphed on a log scale and expressed as cells/mL.

A. Clone 1 has a lower R^* for the limiting resource, and it competitively displaces clone 2.

B. Clone 3 has a lower R^* for the limiting resource and it competitively displaces clone 4.

C. The R^* of clone 4 is unaffected by the presence of a growth inhibitor (nalidixic acid), but the R^* of clone 3 increases with increased concentrations of the inhibitor.

D. By using the concentration of the inhibitor at which the two clones had identical R^*, Hansen and Hubbell demonstrated that two species could coexist at equilibrium on one resource if they had identical requirements for that resource. The growth inhibitor is a "limiting factor." These results thus demonstrate that two species can coexist on one resource and one "limiting factor." See also Figure 90 and Levin (1970).

phosphate at 20°C and both species coexisted for all conditions of silicate and phosphate limitation at this temperature (see Fig. 29). Other examples of two-species competition for one resource, given in Tilman (1981), are also consistent with the theory of resource competition presented here. In all cases, the species with a significantly lower equilibrium requirement

(R^*) for the limiting resource competitively displaced the other species.

The theory of resource competition for a single resource predicts that the species with the lower R^* should be competitively dominant, independent of initial densities. The most powerful test of this prediction would come from the introduction of a single individual of a species with a lower R^* into an equilibrium community limited by a single resource. Just this appears to have happened in some experiments reported by Zevenboom, Van Der Does, Bruning, and Mur (1981). Zevenboom *et al.* were conducting experiments on the light requirements of the nitrogen-fixing blue-green alga, *Aphanizomenon flos-aquae*, by growing a clone of this species in light-limited chemostats. Under such conditions, this species reproduces asexually. One of their cultures reached and maintained an equilibrium for several days, and then showed a dramatic increase in its total biomass. As the total biomass increased, the population shifted from individuals with the normal, wild-type morphology (having heterocysts) to individuals having a mutant (no heterocystic cells) morphology. Zevenboom *et al.* isolated the mutant cells and compared their light requirement with that of the wild-type. As shown in Figure 18, the mutant strain of *Aphanizomenon* has a lower requirement for the limiting resource, light, at the dilution rate used. The ability of a single mutant individual to competitively displace the wild-type individuals is strikingly consistent with the theoretical prediction that the outcome of competition for a single limiting resource should be independent of initial abundances of the competitors. In the context of the discussion of competitive-ability-tradeoffs presented in Chapter 9, it is interesting to note that the mutant individuals, though superior competitors for light, are not capable of fixing atmospheric nitrogen and are thus inferior competitors compared to the wild-type under conditions of limiting nitrogen.

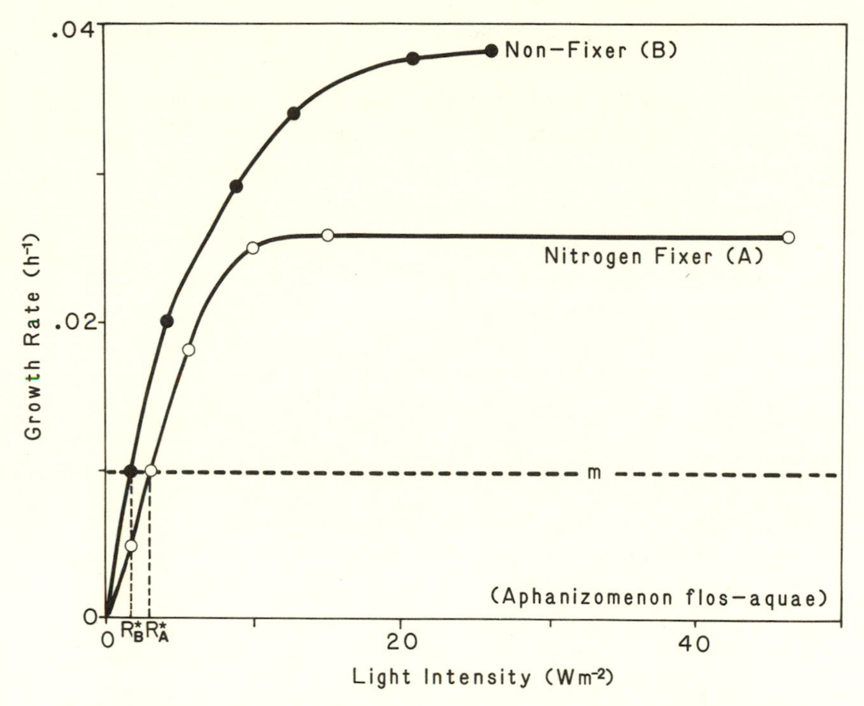

FIGURE 18. Light-dependent growth of the wild type (nitrogen-fixing; labeled *A*) and mutant (non-nitrogen-fixing; labeled *B*) strains of *Aphanizomenon flos-aquae* (from Zevenboom, Van Der Does, Bruning, and Mur, 1981). The mutant type competitively displaced the wild type in a light-limited chemostat for the flow (mortality) rate shown, as resource competition theory would predict, because of its lower R^* for light.

SUMMARY

The theory of equilibrium competition among several species for a single limiting resource presented in this chapter predicts that the one species with the lowest requirement for the resource, as measured by R^*, should competitively displace all other species independent of initial conditions. The experimental work done to date supports this simple prediction. However, many more studies of resource competition among organisms other than algae and bacteria will be needed before the generality of the theory may be determined. The complexity of the life cycle of higher plants and animals suggests that age- or size-dependent resource requirements may be important for such species. The possible complexity added to resource competition by differing age- or size-dependent resource requirements has not yet been explored, either theoretically or experimentally.

CHAPTER FOUR

Competition for Two Resources

This chapter develops the graphical, equilibrium theory of competition for two limiting resources which will be used throughout the rest of this book. The approach developed here is an extension and generalization of work by MacArthur (1972), Maguire (1973), Leon and Tumpson (1975), Petersen (1975), Taylor and Williams (1975), Abrosov (1975), and Tilman (1980). As stated in the generalized equations of consumer-resource interaction offered earlier, four pieces of information are needed to predict the equilibrium outcome of resource competition. These are the reproductive or growth response of each species to the resources, the mortality rate experienced by each species, the supply rate of each resource, and the consumption rate of each resource by each species. An equilibrium occurs when the resource-dependent reproduction of each species exactly balances its mortality and when resource supply exactly balances total resource consumption for each resource.

For the graphical theory developed in this book, the reproductive response is represented by the resource-dependent growth isoclines of each species. The growth isocline along which a species' growth equals its mortality rate is the one at which $dN/dt = 0$, and will be called the Zero Net Growth Isocline, or ZNGI, for that species. Several ZNGI's are shown in Figure 19. For each of these zero net growth isoclines, the population density of the species will increase if a habitat has resource availabilities that are in the shaded region "outside"

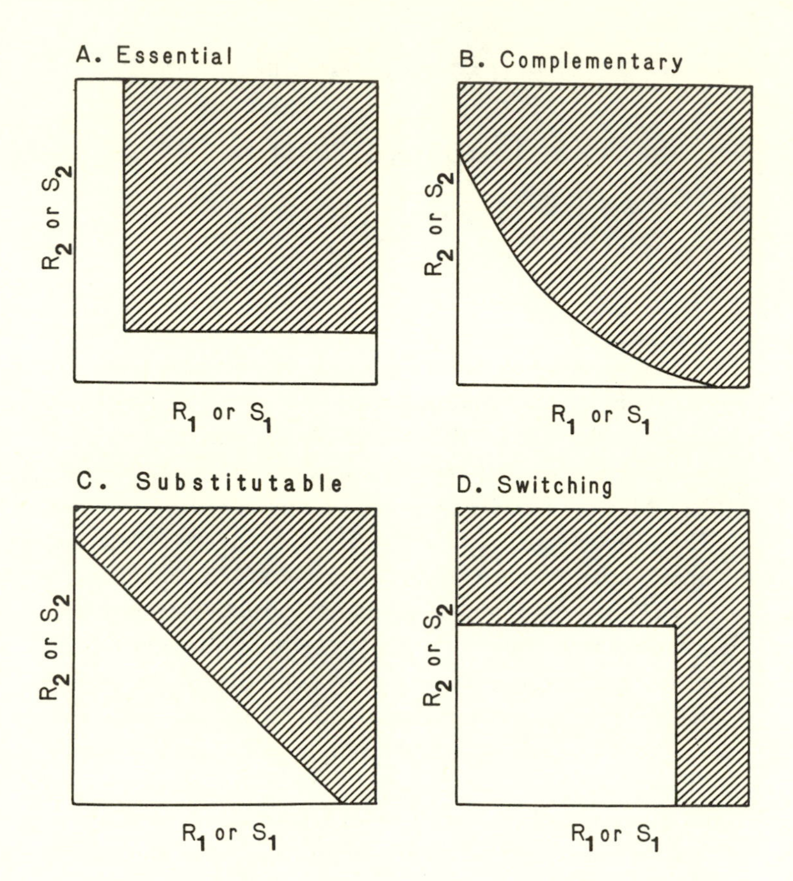

FIGURE 19. The solid curve in each of the figures above is a *z*ero *n*et *g*rowth *i*socline, or ZNGI. The ZNGI represents the amounts of the two resources which must be available in a habitat for a species to maintain an equilibrium population. If a habitat has greater amounts of the resources (the shaded region of these figures), population size will increase. If it has less than the amounts on the ZNGI (the unshaded portion of these figures), population size will decrease. The ZNGI thus shows the amounts of two resources which must exist in a habitat for an equilibrium to occur, i.e., for the resource-dependent reproductive rate to balance exactly the mortality rate experienced by that population.

the growth isocline, and population densities will decrease if resource availabilities fall in the unshaded region "inside" the ZNGI. Population density will remain unchanged only for habitats with resource availabilities on the growth isocline. Thus, the ZNGI specifies half of the information needed to establish an equilibrium.

Also necessary for an equilibrium is the condition that resource consumption rates exactly balance resource supply rates. As briefly discussed in Chapter 2, resource consumption can be represented by a resource consumption vector, the two elements of which give the consumption rates of R_1 and R_2 by this population. As shown in Figure 20, the instantaneous consumption rate of species i of R_1 and R_2 at equilibrium can be represented by a vector which starts at a point on the ZNGI. This vector, labeled $\vec{C}_i$, is the vector sum of two components—the total consumption rates of R_1 and R_2 by this species. Where c_{i1} is the per capita consumption rate of R_1 at equilibrium by species i, c_{i2} is the per capita consumption rate of R_2 at equilibrium by species i, and N_i^* is the equilibrium density of species i,

$$\vec{C}_i = -N_i^* \begin{bmatrix} c_{i1} \\ c_{i2} \end{bmatrix}.$$

Depending on the resource, the values of c_{ij} may be constant or may vary with the position of the equilibrium point on the zero net growth isocline. Both cases are illustrated in Figure 5. This means that the slope of the consumption vector, $\vec{C}_i$, may change for different points along the ZNGI. Note that the slope of the consumption vector is the ratio of the two resources in the diet.

The rate of supply of resources is the final piece of information needed to determine the equilibrium outcome of resource competition. There are many different ways to represent resource supply. The most general definition is to consider

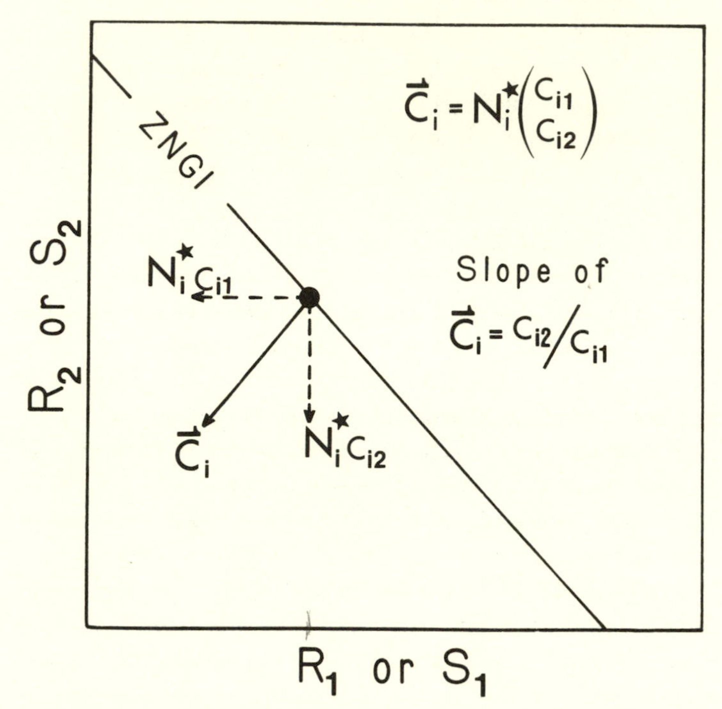

FIGURE 20. The point on the ZNGI is an equilibrium point. The vector that is drawn as a solid line coming from this point is the consumption vector, $\vec{C}_i$. The consumption vector has two components, each representing the consumption rate of one of the resources. These are shown with broken lines coming from the equilibrium point. The slope of the consumption vector is the ratio of these two components.

resource supply in a particular habitat to be represented by a single vector, $\vec{U}$, the resource supply vector. This vector has two components—the supply rate of R_1 and the supply rate of R_2. Such a resource supply vector is shown in Figure 21A, along with its two components. This vector represents the instantaneous rates of supply of R_1 and R_2 for the habitat represented by point A on the ZNGI, i.e., for a habitat that has that particular availability of R_1 and R_2. In Figure 21A, $\vec{U}_A$ is the resource supply vector, U_1 is the supply rate for R_1

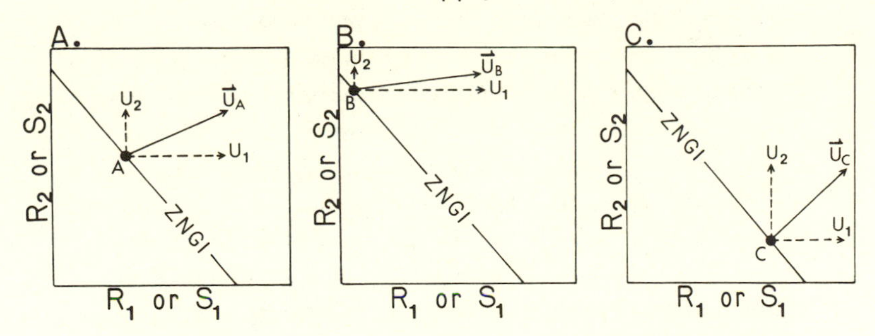

FIGURE 21. Parts A, B, and C of this figure illustrate resource supply vectors, $\vec{U}$, at three different points along a ZNGI. Assuming that the same process is governing resource supply for all parts of this figure, the vectors of parts A, B, and C above demonstrate how the resource supply vector may depend on resource availability. It illustrates the assumption that the rate of supply of a resource decreases as its availability increases.

(and is thus the R_1 component of $\vec{U}_A$), and U_2 is the supply rate of R_2 (and is thus the R_2 component of $\vec{U}_A$).

As illustrated in Figure 21B and C, the resource supply vector is likely to change for different points on the ZNGI. There are several reasons why this may occur. Let us first consider the mineral nutrient resources used by plants. These nutrients occur in the habitat in two forms: as available nutrients and unavailable nutrients. Any given habitat will have a certain total amount of all forms of each nutrient in it. There are numerous physical and biological processes that can change the nutrient from the unavailable to the available form. Assuming that these processes occur at a fixed rate, the rate of supply of a resource would depend on the amount of unavailable resource. The larger the proportion of the total mineral nutrient that is in the unavailable form, the greater would be the supply rate of the nutrient. This assumption of a fixed total amount of all forms of a given nutrient resource in a habitat leads to resource supply vectors that change as illustrated in Figure 21. For instance, habitat B (Fig. 21B) has more available

R_2 than habitat A (Fig. 21A). Assuming the same total amount of all forms of R_2 as in habitat A, this means that habitat B has less R_2 in unavailable forms. Thus, habitat B should have a lower supply rate of R_2 compared to habitat A because there is a smaller pool of unavailable forms from which various physical and biological processes can generate available nutrients. This is illustrated in Figure 21B by having U_2 (the R_2 component of the resource supply vector) be smaller than the U_2 for Figure 21A. Similarly, habitat B has less available R_1 than does habitat A. Thus, the supply rate of R_1 in habitat B should be greater than in habitat A, as illustrated by the larger U_1 of Figure 21B compared to Figure 21A. Similarly, Figure 21C illustrates how the resource supply vector is likely to change if the available amounts of R_1 and R_2 were at point C.

These figures illustrate the assumption that, all else being constant, nutrient supply rates in a given habitat will decrease as the amount of available nutrient in the habitat increases. This generalization seems valid for many other kinds of resources. The assumption of a fixed total amount of all forms of a nutrient in a habitat is somewhat analogous to assuming a fixed "carrying capacity" for that resource in the habitat. The assumption means that there is an upper limit on the amount of available nutrient that can exist in the habitat. All resources, be they nutrients or light for plants or the living resources consumed by herbivores and carnivores, have some upper bound on their density in a particular habitat. Thus, all resources have a "carrying capacity" qualitatively similar to that described for nutrients. Such an upper limit means that the greater the availability of a resource, the lower should be the rate at which it is being supplied to the habitat, as illustrated in Figure 21.

Assuming that the rate of supply of a resource will decrease as the amount of its available form increases seems much more realistic than two alternatives: (1) that supply rate would increase as the amount of the available form increased; or

(2) that supply rate would be unaffected by amount of the available form. If supply rate were to increase as the amount of the available form increased, slight increases in the amount of the available form could lead to a positive feedback loop which would generate resources. If supply rate were independent of availability, the rate of supply of any resource would be constant, with the potential of an infinitely great pool of available resource in the absence of consumption. An assumption of a fixed pool is more realistic. For a living resource, this is comparable to assuming that the resource species has a set upper population density, in the absence of consumption, for a particular habitat, i.e., that it has a "carrying capacity." For ecological reality, it seems necessary to assume that such an upper bound exists for all types of resources.

Let S_j represent the total amount of all forms of resource j which can exist in a habitat in the absence of consumption. S_j is thus somewhat comparable to a "carrying capacity" for that resource. Given this upper bound, a simple approximation for resource supply is to assume that the rate at which the resource becomes available—the supply rate of the resource—is proportional to $(S_j - R_j)$. With this assumption of "density dependent" resource supply, the supply rate of a resource can be approximated by

$$dR_j/dt = a_j(S_j - R_j). \tag{9}$$

For the graphical treatment presented in this book, I will assume that the rate constants, a_j, are equal for all resources. This is a minor assumption, which can easily be changed, and is only made for the simplicity that it gives to graphical analysis.

This leads to a resource supply vector which is defined as

$$\vec{U} = a\begin{bmatrix} S_1 - R_1 \\ S_2 - R_2 \end{bmatrix}, \tag{10}$$

where a is the proportionality rate constant defining resource supply rates.

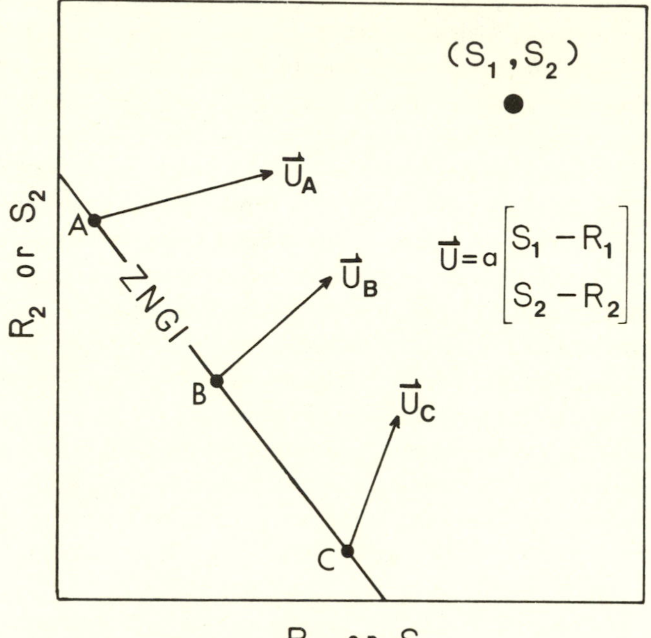

FIGURE 22. Points A, B, and C on the ZNGI represent three habitats which differ in the availability of these two resources, but which have the same resource supply point, (S_1,S_2). The vector labeled $\vec{U}_A$ gives the rate of resource supply for habitat A, and vectors $\vec{U}_B$ and $\vec{U}_C$ show resource supply rates for habitats B and C. Note that for equable resources, the resource supply vector always points toward the resource supply point.

One set of values for a, S_1, and S_2 defines the resource supply rates for a particular habitat. It should be noted that there can be many different resource supply vectors for this habitat, since the resource supply vector is defined both by these three parameters and by R_1 and R_2 (and R_1 and R_2 may be changed by the consumption of resources by the competing species). This is the resource supply process which Stewart and Levin

(1973) called "equable." In Chapter 7 this "equable" mode of resource supply will be compared with a "logistic" mode of resource supply.

The assumptions of Eq. 10 for resource supply can easily be illustrated graphically (Fig. 22). Let the point (S_1,S_2) be called the resource supply point. This point represents the maximal or total amounts of each resource that can exist in a particular habitat. Consider three different habitats, all of which share this same resource supply point and all of which have the same value for a, but which differ in their availabilities of R_1 and R_2. Habitats A, B, and C would have the resource supply vectors shown in Figure 22. Note that for each of these vectors, the vector always points toward the resource supply point, with a length proportional to the distance of the point on the ZNGI from the resource supply point. This simple rule will be applied to all the cases of resource competition considered in this book.

ONE CONSUMER—TWO RESOURCES

The graphical methods just discussed allow the calculation of the equilibrium outcome of consumer-resource interactions. Let us first consider cases in which one species is consuming two resources. Figure 23A shows a case with essential resources. Equilibrium occurs when reproduction and death rates balance, and when resource consumption equals resource supply. Reproduction rate equals death rate for any point on the ZNGI. Of all the points on the ZNGI, there is one point at which the resource consumption vector will be opposite in direction to the resource supply vector. This is the resource equilibrium point, (R_1^*,R_2^*), which is labeled E in Figure 23A. As population density increases, resources will be brought to this point. At this equilibrium point, the population density of the consumer will be such that its instantaneous consumption rate will exactly balance the instantaneous rate of resource supply, as

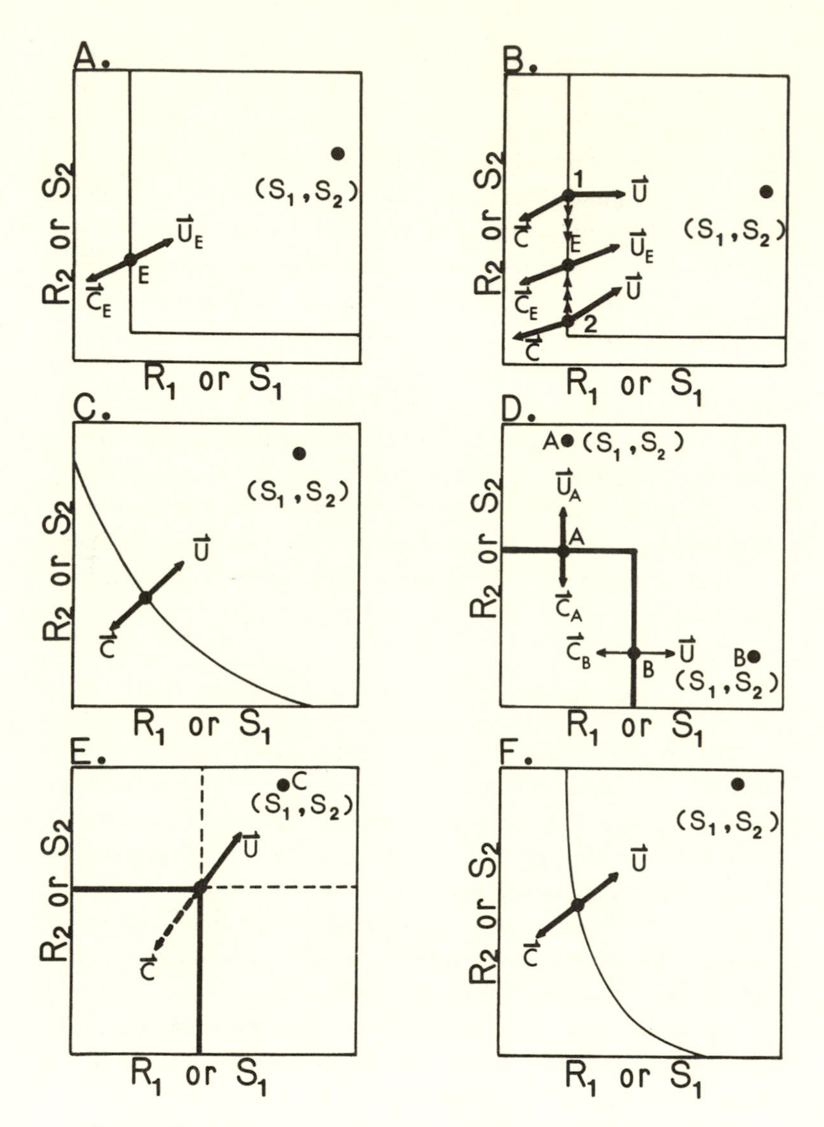

FIGURE 23. This figure illustrates the equilibrium point associated with any given resource supply point, (S_1,S_2), for several different resource classes.

shown. Thus, at equilibrium, the following vector equation will hold:

$$\vec{U} + \vec{C} = \vec{0}. \tag{11}$$

This may be rewritten as

$$a\begin{bmatrix} S_1 - R_1^* \\ S_2 - R_2^* \end{bmatrix} - N^*\begin{bmatrix} c_1 \\ c_2 \end{bmatrix} = \begin{bmatrix} 0 \\ 0 \end{bmatrix}, \tag{12}$$

or $N_i^* = (S_1 - R_1^*)\dfrac{a}{c_1} = (S_2 - R_2^*)\dfrac{a}{c_2}$, where $S_1 > R_1^*$ and $S_2 > R_2^*$.

That such an equilibrium point exists and is locally stable has been demonstrated elsewhere (Tilman, 1980), and I will not pursue this question here. However, the basis for its stability can be seen in Figure 23B. Obviously, the equilibrium point must be on the ZNGI. Figure 23B shows two other points on the ZNGI, one above the equilibrium point (labeled 1), and one below it (labeled 2). For point 1 the consumption vector and the resource supply vector are not exactly opposite in direction. Their sum leads to a net change in resource availabilities. Points on the ZNGI above the equilibrium point have excess consumption of R_2 compared to supply, and the environmental availability of R_2 is decreased. This moves the resource availabilities of this habitat toward the equilibrium point. Below the equilibrium point, there is greater production of R_2 than consumption. The environmental concentration again tends toward that of the equilibrium point. Thus, if a habitat were at resource equilibrium, and were perturbed away from it, the interactions of the consumption and supply processes would tend to *return it to the equilibrium point*. Similarly, if the population density were perturbed from its equilibrium level, it would tend to return, thus suggesting that this equilibrium point is locally stable. However, the equilibrium point, (R_1^*, R_2^*), will move if the position of the ZNGI or of the supply point is changed.

Equilibrium points along the ZNGI for other types of resources are similarly calculated. An example with complementary resources is shown in Figure 23C, and one with hemiessential resources is shown in Figure 23F. In both cases, the equilibrium occurs at that point on the ZNGI at which the resource supply vector is exactly opposite in direction to the consumption vector. This is also the point at which the two vectors have the same slope, i.e., the point at which the ratio of $R_2:R_1$ consumed equals the ratio supplied. Figures 23D and E show an example with switching resources. For supply point A of Figure 23D, the species consumes only R_2, and the equilibrium point is as shown. Similarly, for supply point B, the species consumes only R_1, leading to the equilibrium point shown. However, for supply points within the region between the broken lines of Figure 23E, the species will switch from consuming one to consuming the other resource, always consuming that resource which leads to the greater growth rate. Assuming that the switching is instantaneous, the equilibrium will occur at the kink point of the ZNGI, as shown in Figure 23E, with the species switching from one to the other resource instantaneously as its consumption changes resource availabilities. If switching were to occur with a time lag, resource supply points in the region between the broken lines would lead to sustained resource oscillations around the kink point, with the amplitude of the oscillations dependent on the magnitude of the lag time.

COMPETITION FOR TWO RESOURCES

There are four qualitatively distinct equilibrium cases of resource competition. These are illustrated in Figure 24 using ZNGI for essential resources, although they could be illustrated with growth isoclines for any other type of resource.

Case 1: The ZNGI of species A is always "inside" that of species B. This means that species A requires less of both

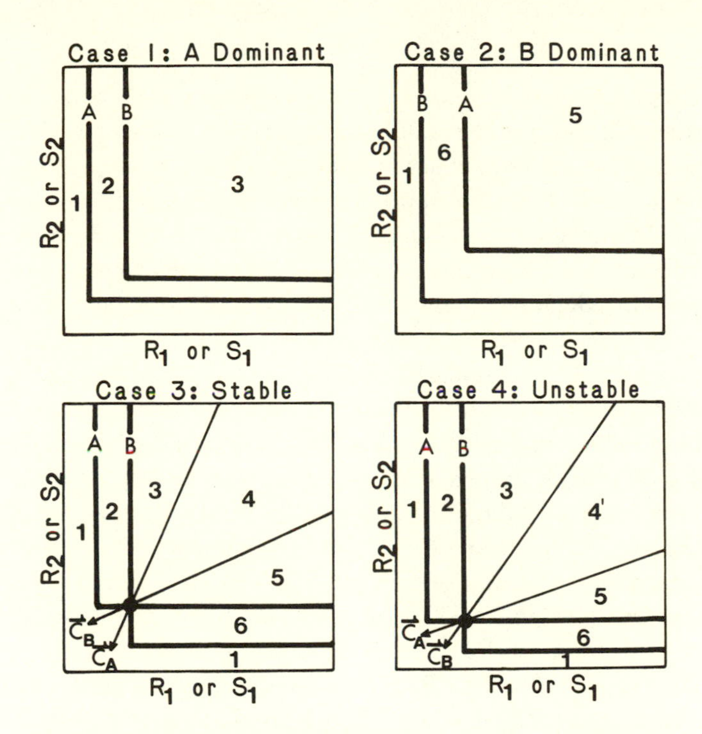

FIGURE 24. The four distinct cases of resource competition are illustrated above. In Case 1, the ZNGI of species *A* is inside that of species *B*. Species *A* will be able to reduce resource levels to a point below that required for the survival of species *B*. Species *A* will competitively displace species *B* for any habitats in which species *A* could survive by itself. Habitats which have a resource supply point, (S_1,S_2), that falls in the region labeled 1 would have insufficient resources to support either species. Habitats with resource supply points in the regions labeled 2 and 3 would be dominated by species *A*, and species *B* would go extinct.

Case 2 is similar to Case 1, except that the ZNGI of species *B* is inside that of species *A*, and species *B* will competitively displace species *A* in regions 5 and 6.

For Case 3, the ZNGI's cross at a two-species equilibrium point, shown with a dot. This equilibrium point is locally stable because each species consumes relatively more of the resource which limits its growth at equilibrium. Habitats which have resource supply points in region 4 will have both species coexisting, while species *A* will dominate in habitats 2 and 3, and species *B* will dominate in habitats 5 and 6.

Case 4 is identical to Case 3 except that the equilibrium point is locally unstable, because each species consumes more of the resource which more limits the other species. The outcomes of competition are identical to those for Case 3, except that in region 4′ either species *A* or species *B* will win, dependent on initial conditions.

resources to grow at a rate equal to its mortality rate than does species *B*. Species *A* should competitively displace species *B* for any conditions in which species *A* can survive (i.e., for any habitats which have resource supply points that fall on or outside the growth isocline of species *A*). There are three different regions in which a resource supply point could fall. Habitats which have resource supply points in region 1 have insufficient resources for the survival of either species. Both species would go extinct. Habitats which have resource supply points in region 2 have sufficient resources for the survival of species *A* but insufficient resources for the survival of species *B*. If these species occurred in such a habitat, species *A* would reach a stable, equilibrium population density and species *B* would go extinct. In region 3, there are sufficient resources for the survival of either species if growing by itself. If both species occurred in a habitat with a resource supply point in region 3, the density of species *A* would increase until resources were lowered to a point on its ZNGI, and species *B* would go extinct.

Case 2: The ZNGI of species *B* is inside that of species *A*. Neither species could survive in habitats with resource supply points in region 1, only species *B* would survive in region 6, and species *B* would competitively displace species *A* in region 5.

Case 3: The ZNGI's of the two species cross. The point at which the isoclines cross is a two-species equilibrium point. This two-species equilibrium point is locally stable. Thus, habitats which have resource supply points that are "focused" into this equilibrium point would have species *A* and *B* coexisting. The resource isoclines and consumption vectors of Case 3 define six regions. Habitats that have resource supply points that fall in region 1 have insufficient resources for the survival of either species. Habitats with supply points in region 2 have sufficient resources for the survival of species *A*, but insufficient resources for the survival of species *B*. Species *A* will dominate this region. Supply points in region 3 have sufficient resources for the survival of either species, if growing alone in such a habitat. If both species occurred in a habitat which had a resource supply point that fell in region 3, species A would competitively displace species B because species A would reduce resource levels down to a point on its ZNGI, but to the left of the two-species equilibrium point, at which level of resource availabilities species B could not survive. Habitats which have resource supply points that fall in region 4 would lead to the stable coexistence of both species. Resource levels would be reduced down to the two-species equilibrium point, at which point each species could grow at a rate exactly equal to its mortality rate. This two-species equilibrium point is locally stable because each species consumes proportionately more of the resource that more limits its own growth (Leon and Tumpson, 1975; Tilman, 1980). This can be seen by noting that at the two-species equilibrium point of Case 3, species A is limited by R_2 and species B is limited by R_1. The consumption vector of species A has a steeper slope than that of species B, indicating that, compared to species B, species A consumes proportionately more of R_2. This may also be stated in terms of nutrient ratios. At equilibrium, species A consumes R_1 and R_2 in the proportion of $R_1 : R_2 = c_{A1}/c_{A2}$. Species B consumes these in the proportion of $R_1 : R_2 = c_{B1}/c_{B2}$. As the vectors are drawn for Case 3, the $R_1 : R_2$ ratio of species A is less than that of species B, indicating that species A consumes proportionately more R_2 than does species B. Its ZNGI shows that species A is limited by R_2 at the two-species equilibrium point. For resource supply points in region 5, species B will competitively displace species A. Only species B would be able to survive for any resource supply points that fell in region 6.

Case 4: As drawn, Case 4 is identical to Case 3 except that the consumption vectors of species A and B have been reversed. Because the consumption vectors have been reversed, the two-species equilibrium point is locally unstable. At equilibrium, species A is limited by R_2 and species B is limited by R_1.

However, the slopes of the consumption vectors show that species A consumes relatively more R_1 than species B. Thus, each species consumes proportionately more of the resource which does not limit its growth. This makes the equilibrium point unstable. The slightest deviation away from the two-species equilibrium point will be magnified through time until one or the other species is competitively displaced. The outcomes of competition for habitats that have resource supply points that fall in regions 1, 2, 3, 5, and 6 are identical to those described for Case 3. However, resource supply points in region 4′ will lead to dominance by either species A or B (with the other species competitively displaced), depending on the initial conditions.

This description of the four general cases of resource competition (Fig. 24) has not yet dealt with what determines the boundary between regions 3, 4, and 5. To determine which habitats will have their availabilities of R_1 and R_2 focused into a two-species equilibrium point by the combined effects of resource consumption and resource supply, it is necessary to consider the dynamics of these processes at the equilibrium point. This may be done using the following vector equation, which defines conditions for which $dR_j/dt = 0$ for all j:

$$\vec{U} + \vec{C}_A + \vec{C}_B = \vec{0}. \tag{13}$$

Written in expanded form this gives the equations

$$a\begin{bmatrix} S_1 - R_1^* \\ S_2 - R_2^* \end{bmatrix} - N_A^*\begin{bmatrix} c_{A1} \\ c_{A2} \end{bmatrix} - N_B^*\begin{bmatrix} c_{B1} \\ c_{B2} \end{bmatrix} = \begin{bmatrix} 0 \\ 0 \end{bmatrix}. \tag{14}$$

This system of equations defines the equilibrium population densities, N_A^* and N_B^*, for any given resource supply point. If the two-species equilibrium point is locally stable, a resource supply point will lead to this point only if the equilibrium densities calculated using this equation are both greater than zero. For N_A^* and N_B^* to be greater than zero, it is necessary that

$$c_{A2}/c_{A1} < (S_2 - R_2^*)/(S_1 - R_1^*) < c_{B2}/c_{B1}, \qquad (15)$$

where c_{ij} is the per capita consumption rate of resource j by species i.

This inequality is easily interpreted graphically. The ratio c_{A2}/c_{A1} is the slope of the consumption vector of species A, and the ratio c_{B2}/c_{B1} is the slope of the consumption vector of species B. Subtracting R_1^* and R_2^* from S_1 and S_2 moves the origin (i.e., the point "zero, zero") up to the two-species equilibrium point. Thus, for a resource supply point to lead to coexistence it must fall between two lines through the equilibrium point: one line having the slope of the consumption vector of species A, and the other line having the slope of the consumption vector of species B. The equilibrium point will be stable if each species consumes proportionately more of the resource that more limits its own growth. This is a mathematical restatement of what was illustrated graphically in Figure 24.

This result may seem deceptively simple. Does it depend on some hidden assumptions of the model? For instance, Figure 5E shows a situation in which a line through the corners of the isoclines for essential resources bends, which means that the optimal ratio of the two resources changes at different growth rates. If this occurs, does the same simple rule specify which resource supply points will lead to coexistence? Several such cases with stable and unstable equilibrium points are illustrated in Figures 25 and 26 from computer simulations of competition for two essential resources. Figure 25A shows ZNGI's for a stable equilibrium point. The graphical theory developed in the preceding pages predicts that resource supply point 1 would lead to dominance by species X and extinction of species Y. The outcome of this case (Case 1) is shown in Figure 25B. Species X reaches an equilibrium population density, as predicted, and species Y is being competitively displaced. Figure

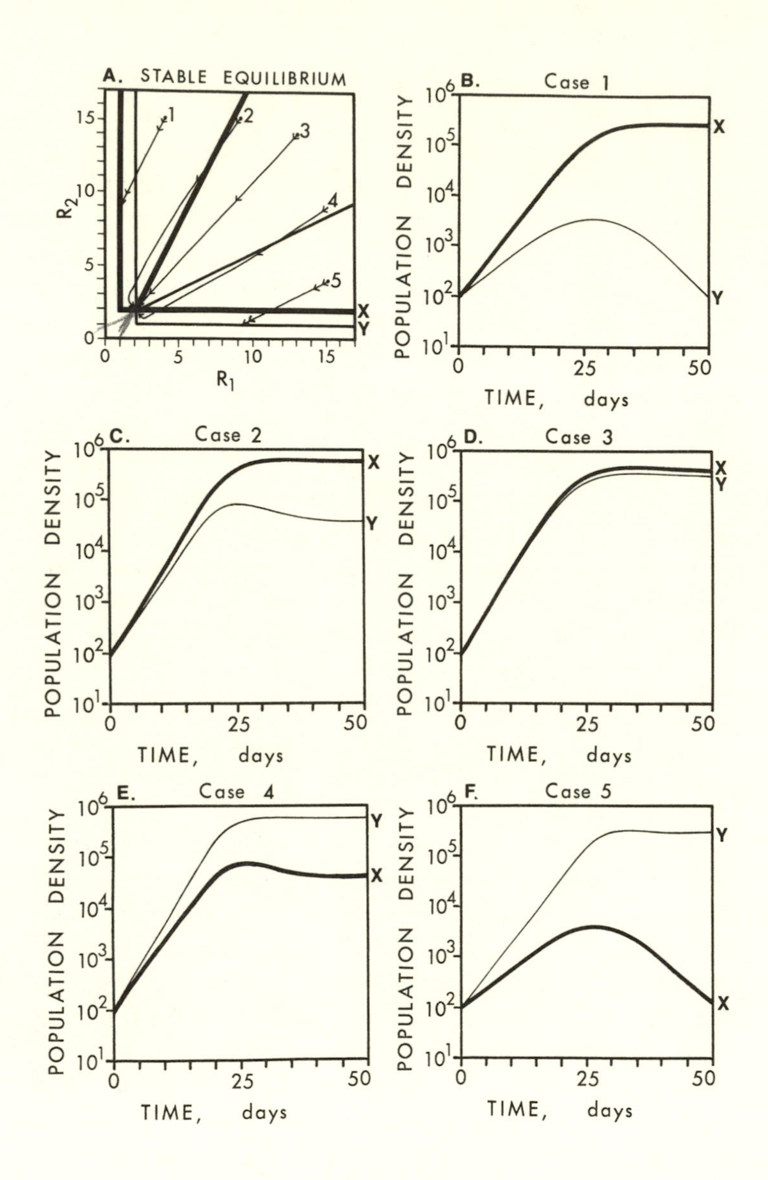

A. STABLE EQUILIBRIUM
R2
R1
1
2
3
4
5
X
Y
B. Case 1
C. Case 2
D. Case 3
E. Case 4
F. Case 5
POPULATION DENSITY
TIME, days

Figure 25. A. The outcome of competition for five different resource supply points is illustrated for a situation with a stable equilibrium point. The five resource supply points used are numbered 1 to 5 in this part of the figure, and called Cases 1 to 5 in the other parts of the figure. The vectors coming from the resource supply points showthe dynamics of resource availability. In all cases, the availabilities of the two resources are reduced down to a point on a ZNGI.

B to F. These graphs give the dynamics of population change for these resource supply points, as determined by solution of a Monod model of competition for two essential resources. The outcome of competition among these two species as determined by numerical solution of these Monod competition equations is always consistent with the outcome predicted by the graphical theory developed in this chapter. Supply points 2 and 4 illustrate that a resource supply point which is within the region of predicted coexistence does lead to stable coexistence even if the interaction of consumption and supply processes temporarily takes the resource availabilities outside the region of stable coexistence.

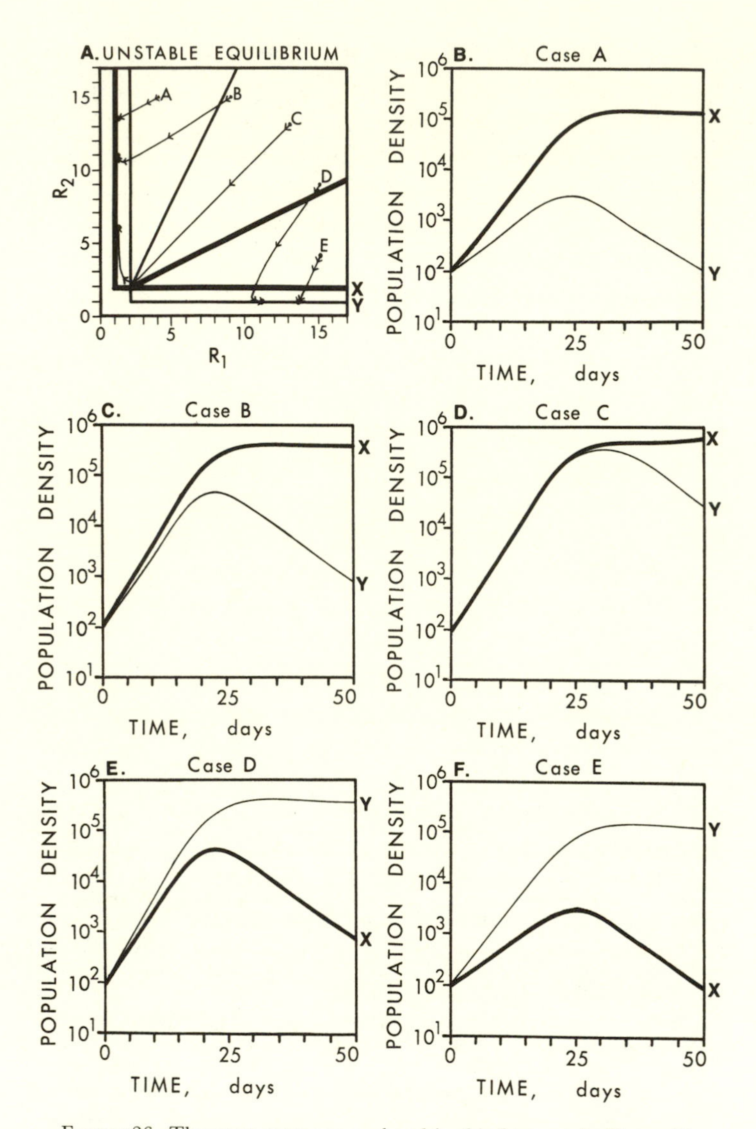

FIGURE 26. The same cases are explored in this figure as in Figure 25, except that the two-species equilibrium point is locally unstable. Because it is locally unstable, either species X or Y will competitively displace the other species, at equilibrium, for any supply point in the region bounded by the consumption vectors coming from the equilibrium point.

25A also illustrates the changes in availabilities of R_1 and R_2 by showing the trajectory of resources, starting at the resource supply point. Note, for Case 1, that resources are consumed down to a point on the ZNGI of species *X*, as the graphical approach predicted. Resource supply point 5 and Case 5 (Fig. 25F) similarly illustrate a situation in which species *Y* competitively displaces species *X*, as predicted. Supply points 2, 3, and 4 are all in the region in which the graphical approach predicts stable, equilibrium coexistence. As illustrated in Figure 25C, D, and E, these two species do stably coexist for these points. However, as shown in Figure 25A, the trajectories of resource availability are curved. In Cases 2 and 4, this takes the habitat's resource availabilities temporarily out of the region of coexistence, but the joint processes of consumption and supply eventually lead to the two-species equilibrium point and to stable coexistence. Thus, the long-term, equilibrium outcome of competition is identical to that predicted by the simple graphical approach, even though the short-term dynamics can lead to a complex resource availability trajectory. In all cases, the equilibrium outcome of competition is determined by the region in which the resource supply point falls, independent of initial resource availabilities, initial population densities, and curved resource trajectories. The long-term effect of the interaction between resource supply and consumption is to focus the resource availabilities into the stable, two-species equilibrium point for any habitat which has a resource supply point that falls in the region of coexistence.

This is further illustrated in Figure 26, in which the two-species equilibrium point is locally unstable. Again, the resource availability trajectories and population dynamics are consistent with the simple graphical theory. Resource supply point *A* is in the region in which species *X* should be dominant, as it is. The habitat's resource availabilities end up at a point on the ZNGI of species *X*. For resource supply points *B*, *C*, and *D*, theory predicts that either species *X* or *Y* should win, dependent on initial conditions. For point *B*, species *X* wins because of its ability to drive resource availabilities into the region in which it is a superior competitor. Resource supply point *C* is right on the dividing line between dominance by species *X* and dominance by species *Y* when the species start with equal densities. Resource levels are driven down to the two-species equilibrium point, from which they deviate away toward dominance by species *X*. For point *D*, species *Y* dominates, as it does for point *E*. For both stable and unstable two-species equilibrium points, the long-term outcome of competition is consistent with the simple rules illustrated in Figure 24.

These results allow a simple, graphical treatment of resource competition. Several cases are illustrated in Figure 27, using the consumption vectors derived in Chapter 2 (Fig. 5). In each case, two species are shown competing for two resources, and isoclines have been chosen so that they cross. For each case shown, the two-species equilibrium point is locally stable because each species consumes more of the resource that more limits its own growth rate. As long as their isoclines cross, it is possible for two species to coexist stably on two resources. The growth isoclines and consumption vectors of the species define the habitats (resource supply points) in which two species would stably coexist, one or the other species would be competitively dominant, or neither species would survive.

Figure 27 suggests several interesting relationships. First, it suggests that stable, two-species equilibrium points may be the norm for consumer-resource interactions, independent of the type of resource for which competition occurs. The stability of the equilibrium points of Figure 27 comes from the consumption vectors used. These consumption vectors were derived in Chapter 2 using a simple theory of optimal foraging. Thus, if foraging is optimized as suggested in Figures 3 to 9, two-species equilibrium points should be locally stable. This suggests that, in an evolutionary sense, stable coexistence of various pairs of species can be considered to be an incidental effect of selection

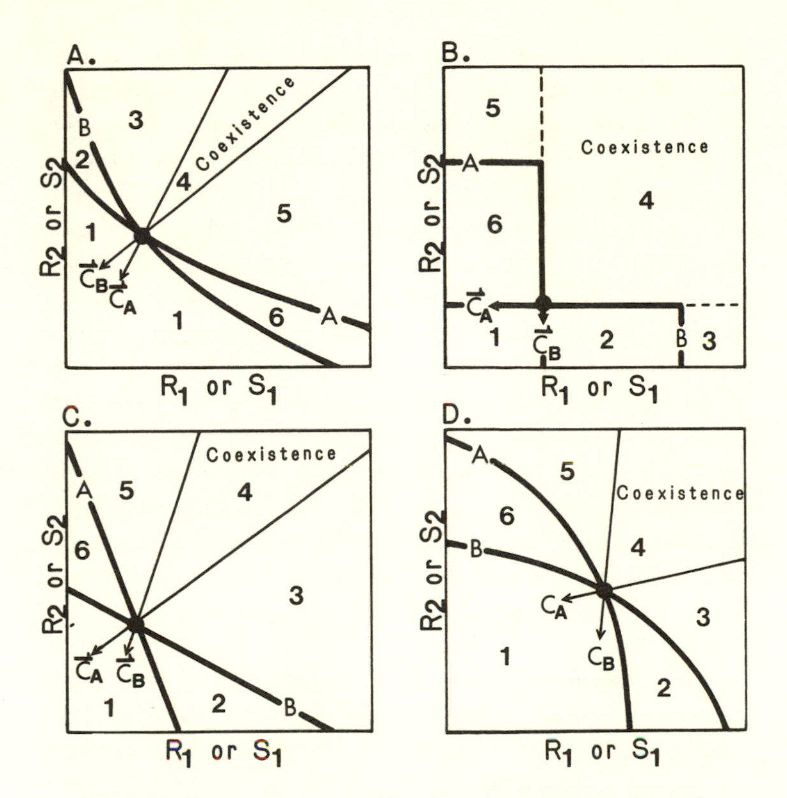

FIGURE 27. Four cases of competition for two resources are shown. In each case, the ZNGI's cross at a stable equilibrium point. The predicted outcome of competition for habitats that have resource supply points that fall in the regions numbered from 1 to 6 is: region 1—extinction of both species; region 2—dominance by species *A* with species *B* unable to survive even in the absence of species *A*; region 3—species *A* competitively displaces species *B*; region 4—stable coexistence of both species; region 5—species *B* competitively displaces species *A*; region 6—species *B* is dominant because it is able to survive for resource supply points in this region whereas species *A* is unable to survive even in the absence of competition.

for optimal foraging acting on the individuals of each species. Secondly, these results suggest that the type of resources for which competition occurs will not lead to major, qualitative differences in the ecological patterns that can result from competition between two species for two resources. The same qualitative outcomes are possible in all cases: there is a region in which neither species can survive, regions in which one species will be competitively dominant, and a region in which both species will stably coexist. The regions are arranged in a qualitatively similar manner for each of the cases of Figure 27 and for Figure 25. This might seem to imply that the structure of plant communities (which consume essential resources) should be qualitatively similar to that of motile animals (which consume substitutable resources). However, as discussed in the final chapter of this book, the similarities that appear in cases of competition between two species for two resources vanish when cases of competition among numerous species are considered.

All of the cases of competition between two species for two resources discussed so far have assumed that both species are responding to the resources in the same manner. This need not be the case, especially for resources which are nutritionally substitutable. As discussed in Chapter 2, the shape of the resource-dependent growth isocline of a species which is consuming nutritionally substitutable resources depends critically on the relative shapes of the nutrition isocline and the consumption constraint curve (see Figures 3 to 9). The consumption constraint curve reflects the limitations placed on consumption both by the habitat structure and by the foraging methods of the consumer. Similarly, the shape of the nutrition isocline is determined by the physiology of the particular consumer species. Thus, it seems likely that two species which are consuming the same two resources in the same habitat could respond to them quite differently. Several such cases are illustrated in Figure 28.

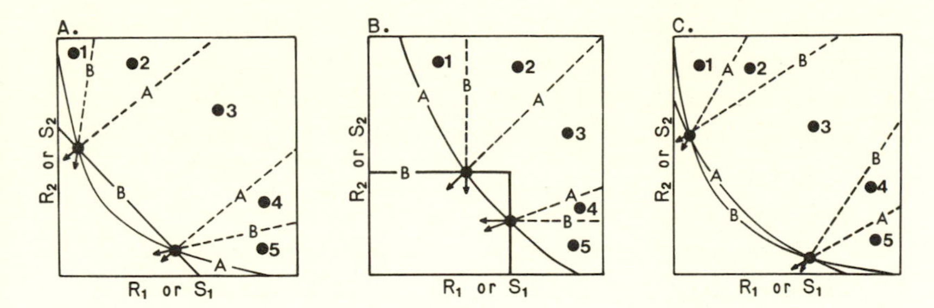

FIGURE 28. Competition between two species for two limiting resources can result in more than a single two-species equilibrium point, as illustrated in the three graphs above.

A case of competition between a species responding to two resources as perfectly substitutable (species B) and a species responding to them as complementary (species A) is illustrated in Figure 28A. Note that there are two two-species equilibrium points, and that these are locally stable. The presence of two equilibrium points leads to a different pattern of dominance of species than is observed for cases of competition for the same type of resources. For Figure 28A, species B should be competitively dominant in habitats which are low in R_1 but high in R_2, such as at resource supply point 1. Species A and B should coexist for habitats which are slightly more rich in R_1 but more poor in R_2, such as at supply point 2. Habitats which are even more rich in R_1 but poorer in R_2 would be dominated by species A (resource supply point 3). Further changes in the relative abundances of these resources in this same direction would lead to resource supply point 4, at which both species stably coexist, and then to point 5, at which species B is dominant. If a series of habitats were arranged from those most rich in R_2 but also most poor in R_1 to those most poor in R_2 but most rich in R_1, this would give a "resource ratio gradient." For the gradient from point 1 to point 5 in Figure 28A, species B is competitively dominant at both ends, and species A is competitively dominant in the middle. This pattern is qualitatively different from the patterns predicted for any of the cases of Figures 25 and 27, in which one species is dominant at one end of the gradient, the two species coexist in the middle of the gradient, and the other species is dominant on the other end of the gradient.

A case in which one species responds to two resources in a switching manner and the other responds to them as complementary resources is illustrated in Figure 28B. Again, there are two stable two-species equilibrium points. The pattern of dominance of these species along a resource gradient is similar to that for Figure 28A. Figure 28C shows a case in which both species respond to the resources as complementary, but the complementarity is much greater for species *B* than for species *A*. This can lead to two two-species equilibrium points, and to a pattern of dominance along a resource ratio gradient as described for Figure 28A. There are many other cases in which competition for different classes of resources or competition for the same class of resources can lead to two or more two-species equilibrium points, which may be stable or unstable, as sketching curves would easily show. However, there can be at most one two-species equilibrium point for cases of two species competing for essential or for switching resources, as dictated by the geometry of the ZNGI's. In this way, essential and switching resources are qualitatively different from all other resource types.

EXPERIMENTAL STUDIES OF RESOURCE COMPETITION

There have been several experimental studies of competition for two resources which have been sufficiently complete to allow them to be analyzed using the theory developed in this book. I will go through these in detail, illustrating with experimental results the concepts presented up to this point.

In 1976 and 1977, I published a series of papers (Titman, 1976; Tilman and Kilham, 1976; Tilman, 1977) dealing with

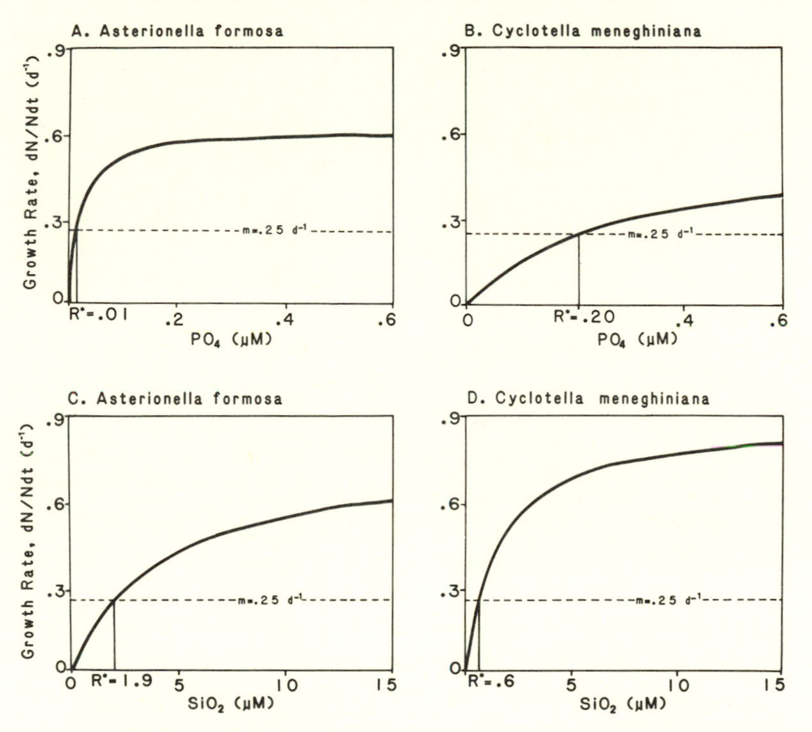

FIGURE 29. A and B. The experimentally determined requirements of two species of freshwater algae (*Asterionella formosa* and *Cyclotella meneghiniana*) for limiting phosphate (from Tilman and Kilham, 1976).

C and D. The requirements of these same species for limiting silicate. The predicted requirements of these species for these two resources at a mortality rate of 0.25 day^{-1} are labeled R^*.

competition between two species of freshwater diatoms for various proportions of limiting silicate and phosphate. The phosphate-limited growth curves for these two species, *Asterionella formosa* and *Cyclotella meneghiniana*, are shown in Figure 29A and B, and the silicate limited growth curves in Figure 29C and D. This information was then used to predict the outcome of semicontinuous "chemostat" competition experiments for limiting silicate and phosphate for a mortality (dilution) rate of 0.25 day^{-1}.

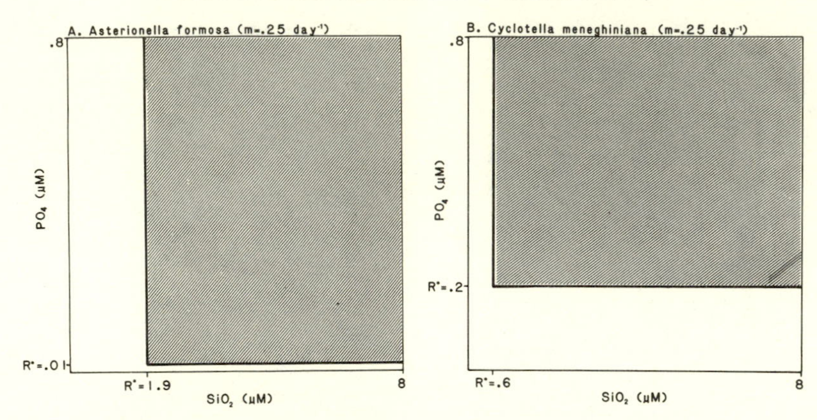

FIGURE 30. A and B. The data of Figure 29 were used to predict the position of the ZNGI of each of these species. Note that the position of the ZNGI for essential resources is determined by the R^* values for each species. Each species should increase for resource availabilities in the shaded region of the figure, and should decrease for resource availabilities in the unshaded region of each figure.

The first step in such predictions is determination of the position of the ZNGI for each species. For essential resources, the position of the ZNGI is determined by the R^* of each species for each of the limiting resources. As calculated from the data of Figure 29, *Asterionella* requires at least 1.9 μM of silicate and at least 0.01 μM of phosphate in order to maintain a stable population when it experiences a mortality rate of 0.25 day^{-1}. These requirements define the ZNGI of *Asterionella* shown in Figure 30A. Similarly, *Cyclotella* requires at least 0.6 μM of silicate and 0.2 μM of phosphate to grow at a rate of 0.25 day^{-1}. Its ZNGI is shown in Figure 30B. Each species can increase in population density for any habitat which has a resource supply point that falls in the shaded region; population density would be constant for a resource supply point on the ZNGI, and decrease in the unshaded region of Figure 30. The superimposed isoclines, the consumption vectors, and the predicted outcomes of competition are shown in Figure 31A, along with the outcome of numerous competition experiments.

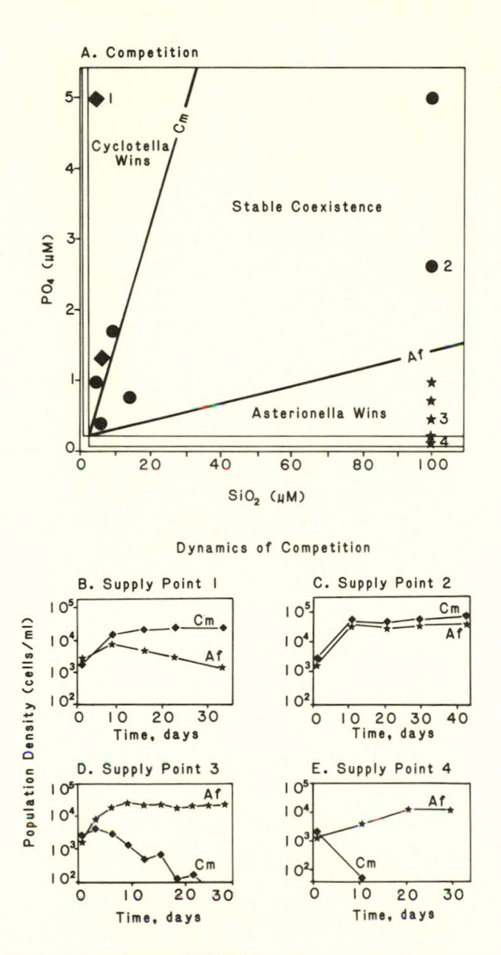

FIGURE 31. A. The observed ZNGI's and consumption vectors of each species were used to predict the outcome of silicate and phosphate competition between these two species. The results of competition experiments are shown with stars for experiments in which *Asterionella* was competitively dominant, with dots for experiments in which the species coexisted, and with diamonds when *Cyclotella* was dominant.

B to E. The experimentally observed dynamics of competition are illustrated for the four resource supply points numbered in part A of this figure. (From Tilman, 1977.)

The resource supply points that led to dominance by *Asterionella* are shown with a star, those that led to dominance by *Cyclotella* are shown with a dot, and those that led to coexistence of both species are shown with a square in Figure 31A. There is close agreement between the actual and predicted results. The dynamics of nutrient competition for four of these cases are shown in parts B, C, D, and E of Figure 31. Part B shows the observed dominance of *Cyclotella* for the resource supply point numbered 1 in Figure 31A, for which both species are silicate limited. Part C shows the coexistence of these species for resource supply point 2. Parts D and E show competitive displacement of *Cyclotella* by *Asterionella* under conditions in which both species are phosphate limited (resource supply points 3 and 4). In all but 2 of the 13 competition experiments at a mortality rate of 0.25 day^{-1} (Tilman, 1977), the long-term outcome of competition was consistent with the predictions of the graphical, equilibrium theory. The two exceptions were very close to the predicted boundary between dominance and coexistence.

The work of Holm and Armstrong (1981) illustrates an interesting case of competition for two limiting resources. They studied the effects of limiting silicate and phosphate on competition between a diatom and a blue-green alga. Although diatoms require both of these resources for growth, blue-green algae do not require any silicate for normal growth. Thus, the shapes of the isoclines for these two species differ. Using the data reported in the Holm and Armstrong paper, it is possible to calculate the position of the isoclines for these two species. For competition in flow-through culture at a dilution rate of 0.11 day^{-1}, their clone of *Asterionella* would require at least 0.41 μM of silicate and 0.01 μM of phosphate, whereas the blue-green alga *Microcystis aeruginosa* would require no silicate but 0.13 μM of phosphate to grow at this rate. This leads to the isoclines of Figure 32A. Notice the large region of predicted coexistence, caused by the lack of a silicate requirement by

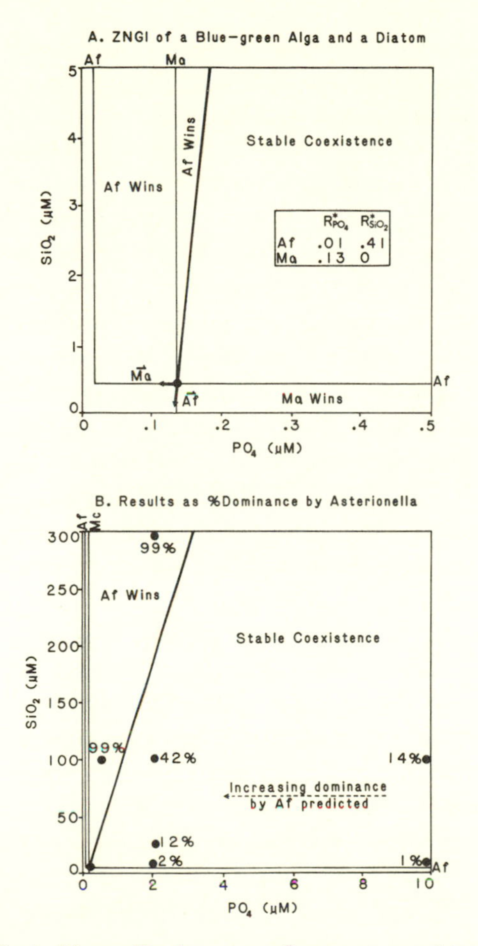

FIGURE 32. A. The predicted outcome of resource competition between a blue-green alga (*Microcystis aeruginosa*; Ma) and a diatom (*Asterionella formosa*; Af), both isolated from Lake Michigan. The isoclines and consumption vectors shown were derived from the data presented in Holm and Armstrong (1981).

B. The predicted outcomes of competition can be compared with the outcomes observed by Holm and Armstrong (1981). The observed percent relative abundance of *Asterionella* in the competition flasks, calculated on a biomass (biovolume) basis, is shown for the competition experiments performed at this flow (mortality) rate. For the supply points in the region in which *Asterionella* is predicted to competitively displace *Microcystis*, *Asterionella* is 99% of the total biomass. In the region of predicted coexistence, *Asterionella* ranges from 1 to 42% of the biomass, in a manner generally consistent with the predictions of theory.

blue-green algae. The results of their competition experiments are summarized in Figure 32B. Each point shows a resource supply point that Holm and Armstrong used for a competition experiment. The number written next to the point is the percent abundance of *Asterionella* at the termination of the experiment, calculated in terms of biomass. Both species stably coexisted in the predicted region of coexistence. The percent dominance of *Asterionella* increased with increases in the supply rate of silicate, as expected. In the region in which *Asterionella* was predicted to competitively displace *Microcystis*, *Asterionella* was always at least 99% of the total biomass, indicating that it was competitively dominant in that region. However, *Microcystis* did maintain an apparently stable background population of less than 1% of the total biomass even under these conditions, and it was not completely competitively displaced, as an equilibrium theory would predict. Holm and Armstrong (1981) have suggested several possible reasons for this, including the tendency for *Microcystis* to adhere to the walls of the culture vessel. Whatever accounts for the very small background population of *Microcystis* for resource supply points for which *Asterionella* should be competitively dominant, the basic pattern observed in these experiments is consistent with theoretical predictions.

Some other cases of two species competing for two resources come from Tilman (1981). Again, the competing species were isolated from Lake Michigan, and their silicate and phosphate requirements used to determine the position of their ZNGI's and the slope and length of their consumption vectors. Phosphate and silicate were chosen because they are commonly limiting resources in Lake Michigan. Figure 33 shows a case of competition in which the ZNGI of one species (*Tabellaria fenestrata*) is outside that of the other species (*Fragilaria crotonensis*), as illustrated for Case 1 of Figure 24. For each of the three resource supply points used in competition experiments, *Fragilaria* competitively displaced *Tabellaria*, as predicted by

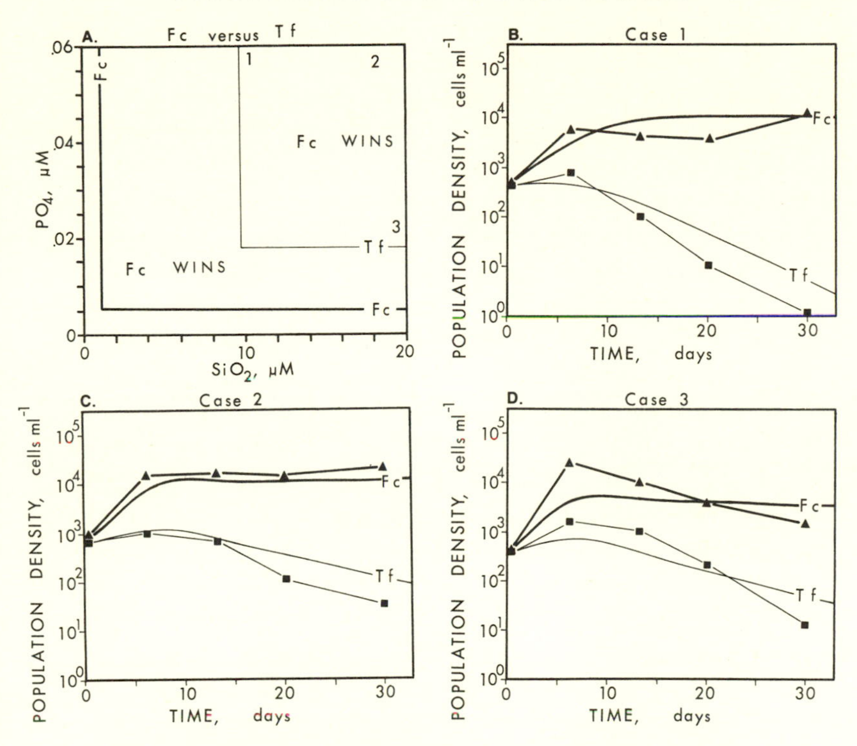

FIGURE 33. Competition between *Fragilaria* and *Tabellaria* (freshwater diatoms) for phosphate and silicate (from Tilman, 1981). As shown in part A, the ZNGI of *Tabellaria* (Tf) is always outside that of *Fragilaria* (Fc), and it should always be competitively displaced by *Fragilaria*. For the three supply points used, this result was observed (parts B, C, and D). See also Figures 52 and 54.

theory. Figure 34B-D shows cases of competition between two species (*Fragilaria crotonensis* and *Asterionella formosa*) which single-species, resource-limited growth experiments predicted to have almost identical requirements for silicate and phosphate. As predicted by theory, within the time span of the experiments neither species was competitively displaced, with both species having almost identical population dynamics. Several other cases of two-species competition for two resources

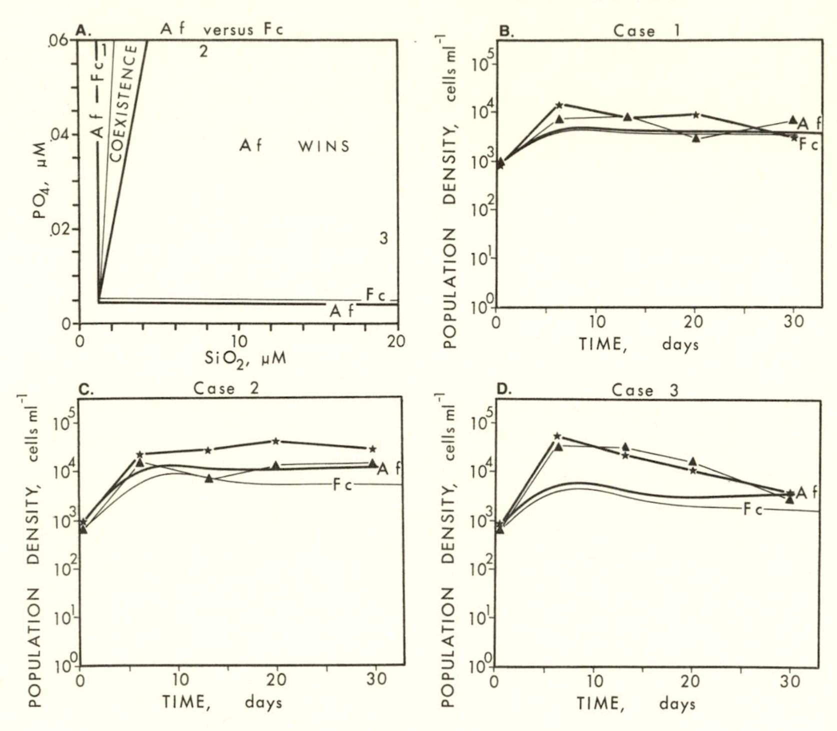

FIGURE 34. *Fragilaria* (Fc) and *Asterionella* (Af) were predicted to be essentially identical in their requirements for silicate and phosphate at 20°C (part A above, from Tilman, 1981). These species apparently coexisted for the three resource supply points used, as shown in parts B, C, and D, above.

were given in Tilman (1981). None is inconsistent with the equilibrium predictions of resource competition theory. However, the similar phosphate requirements of several species made some cases "too close to call."

A tentative example of competition between two species of terrestrial plants for two nutrients comes from Braakhekke's (1980) study of *Plantago lanceolata* and *Chrysanthemum leucanthemum* competing for potassium and calcium. As discussed in Chapter 3, *Plantago* has a lower requirement for calcium and is the superior competitor when both species are calcium limited. Calculations based on Braakhekke's data for potassium-limited monocultures of each species suggest that both species can reduce potassium to indistinguishably low levels. Because of *Plantago*'s lower calcium requirement, a two-species equilibrium point could only exist if *Chrysanthemum* had a lower potassium requirement than *Plantago*. If there were such a two-species equilibrium point, it would occur with *Plantago* being potassium-limited and *Chrysanthemum* being calcium-limited. Braakhekke's studies of calcium and potassium consumption by these species, however, indicate that, compared to *Chrysanthemum*, *Plantago* consumed relatively more calcium than potassium. This means that a two-species equilibrium point, if it were to exist, would be unstable. These results thus predict that there should be no ratios of calcium to potassium at which these two species could stably coexist. This qualitative prediction is consistent with the results of Braakhekke's experiments: the two species did not coexist at any of the three Ca:K ratios used. Rather, *Plantago* was the superior competitor in all the experiments. The superior competitive ability of *Plantago* under conditions in which both species were probably potassium limited suggests that it may be a superior competitor for both resources compared to *Chrysanthemum*. If further work indicates that this is the case, competition between these species for Ca and K would be similar to competition between *Fragilaria* and *Tabellaria* for P and Si, as illustrated in Figure 33.

Braakhekke's results raise an interesting question. Even if a two-species equilibrium point does exist between these two species, the equilibrium point should be locally unstable. How, then, could these species coexist in nature, as they are so often observed to do? I do not have an answer to this question, but I do have a suggestion as to how the answer might be found. Experimental work on competition between these two species in the field (Van den Bergh and Elberse, 1975; Van den Bergh and Braakhekke, 1978; Braakhekke, 1980) strongly suggested that light and nitrogen were the two most important limiting resources. If resource competition is responsible for the coexistence of these species in the field, only those resources which are actually demonstrably limiting in the field need (or should) be studied. If additions of calcium and potassium indicate that these are not limiting resources for these species in their natural habitat, the responses of these species to Ca and K are irrelevant to the question of their natural coexistence. Resource competition theory predicts that naturally coexisting species should have differing requirements for limiting resources, but places no constraints on the requirements of species for non-limiting resources. It is imperative that ecological studies of resource competition focus on those resources which actually limit species in their natural habitats, if the results of the experiments are to be used to interpret patterns in the natural world. Although studies of competition for resources which are never limiting in nature have some short-term heuristic value, in the long term such studies may confuse much more than they clarify.

SUMMARY

This chapter has shown how resource requirements of species, represented by resource-dependent growth isoclines, may be used to predict the equilibrium outcome of competition between two species for two resources (Fig. 24). If the zero net growth isoclines (ZNGI's) of two species do not cross, the species with the lower requirement for the resources should competitively displace the other species for all starting conditions. If the isoclines cross, the point at which they cross is either a stable or an unstable two-species equilibrium point. It will be a point of stable coexistence of two species if, at that equilibrium point, each species consumes proportionately more of the resource that more limits its growth. The equilibrium will be unstable if each species consumes proportionately more of the resource that less limits its own growth rate. Several experimental studies of two-species competition for two resources are consistent with the graphical, equilibrium theory of resource competition presented. However, the available experiments are all for freshwater algae. Many more tests using animals and other plants are needed.

CHAPTER FIVE

Spatial Heterogeneity, Resource Richness and Species Diversity

Several recent papers (e.g., Grubb, 1977; Connell, 1978; Huston, 1979) have argued that only non-equilibrium approaches can explain the patterns of species diversity observed in plant communities. Grubb asserted that the existence of "a million or so animals can easily be explained in terms of the 300,000 species of plants (so many of which have markedly different parts such as leaves, bark, wood, roots, etc.), and the existence of three to four tiers of carnivores (Hutchinson, 1959)," but that "there is no comparable explanation for autotrophic plants; they all need light, carbon dioxide, water and the same mineral nutrients." For similar reasons, Connell (1978) and Huston (1979) assert that the "niche diversification hypothesis" is not applicable to plant communities, and suggest, instead, that continual, small-scale disturbances, which prevent competitive displacement, maintain the diversity of plant communities. This view may be contrasted with the approach developed in this chapter which suggests that patterns of species richness in plant communities may be explained by an equilibrium theory of plant competition for limiting resources. This analysis is not offered as a refutation of the possible importance of periodic disturbances in communities of sessile plants and animals, which is discussed in Chapter 8, but as an alternative hypothesis which seems to be supported by numerous correlational and experimental data.

The preceding chapter presented a theory of competition between two species for two limiting resources, and derived rules which predicted the equilibrium outcome of competition in a spatially homogeneous habitat (i.e., for any given resource supply point). In this chapter, that theory is extended to communities of numerous species competing for two resources in spatially heterogeneous habitats. The approach thus developed is then used to explore the theoretical effects of habitat resource richness and habitat enrichment (fertilization) on community structure. This theoretical analysis is followed by a brief review of the observed relationships between resource richness and plant community diversity.

This analysis, which is centered on terrestrial and aquatic plant communities, was begun because of the numerous lines of evidence linking changes in nutrient resources with changes in the species composition and diversity of plant communities. The major pattern which led to this work has been repeatedly observed in a wide variety of plant communities, and has been termed the "paradox of enrichment" by Rosenzweig (1971). In aquatic and terrestrial habitats, nutrient enrichment leads to decreased species diversity, with the type of enrichment influencing which species become dominant (e.g., Lawes *et al.*, 1882; Milton, 1947; Thurston, 1969; Willis, 1963; Schindler, 1977). This is one of four lines of evidence which suggest that competition for inorganic nutrients influences plant community structure. A second piece of evidence, and a prerequisite condition, is that in their various mineral forms N, P, K, Mg, Ca, and about fifteen other elements are universally required for plant growth (Salisbury and Ross, 1969), with the reproductive rate of a plant depending on the concentration of its limiting nutrient (e.g., Bradshaw *et al.*, 1958, 1960a, 1960b, 1964; Rorison, 1958; Phares, 1971; and Moore *et al.*, 1973 for terrestrial plants; and Droop, 1974; Tilman and Kilham, 1976; and Rhee, 1978 for aquatic plants). Third, nutrient addition experiments in aquatic and terrestrial habitats have indicated that almost all plants are nutrient limited, in that their growth (reproductive) rate increases with fertilization (e.g., Thurston, 1969; Bradshaw, 1969; Ellis, 1971; Powers *et al.*, 1972). Fourth, the spatial or temporal distribution of many species is often closely correlated with the spatial distribution of plant nutrients (e.g., Snaydon, 1962; Pigott and Taylor, 1964; Lund *et al.*, 1963; Zedler and Zedler, 1969; Hanawalt and Whittaker, 1976). Bradshaw (1969) concluded that the requirements of all plants for mineral nutrients, the stimulation of plant growth by nutrient addition in the field, and the changes observed in plant communities after fertilization strongly suggest that competition for nutrients is the major factor determining the species composition of natural plant communities.

In order to explore the implications of resource competition for natural plant communities it is necessary to consider the process of nutrient competition in these communities. Hulburt (1970) and Harper (1977) have advanced similar ideas on the processes of nutrient competition in aquatic and terrestrial communities, respectively, suggesting that competition can occur only between neighboring individuals. In order to apply the theory developed in this book to competition between neighboring individuals, it is necessary to know the resource supply rates *experienced by these individuals*. It is very unlikely that each plant in a given habitat will experience the same rate of supply of each resource. For terrestrial plants, small-scale variations in soil type, topography, soil moisture content, and slope should cause individual plants in a habitat to experience different average rates of nutrient supply. A feeling for the spatial complexity of soils may be gained from Figure 35, which shows contours of total soil nitrogen (TN), extractable magnesium (Mg), and contours of the ratio of total soil nitrogen to extractable magnesium (TN : Mg). These soil nutrient maps are for a 12 × 12 meter area of an agricultural field at Cedar Creek Natural History Area, Minnesota, which had been abandoned 10 years previous to sampling. Nitrogen and magnesium were

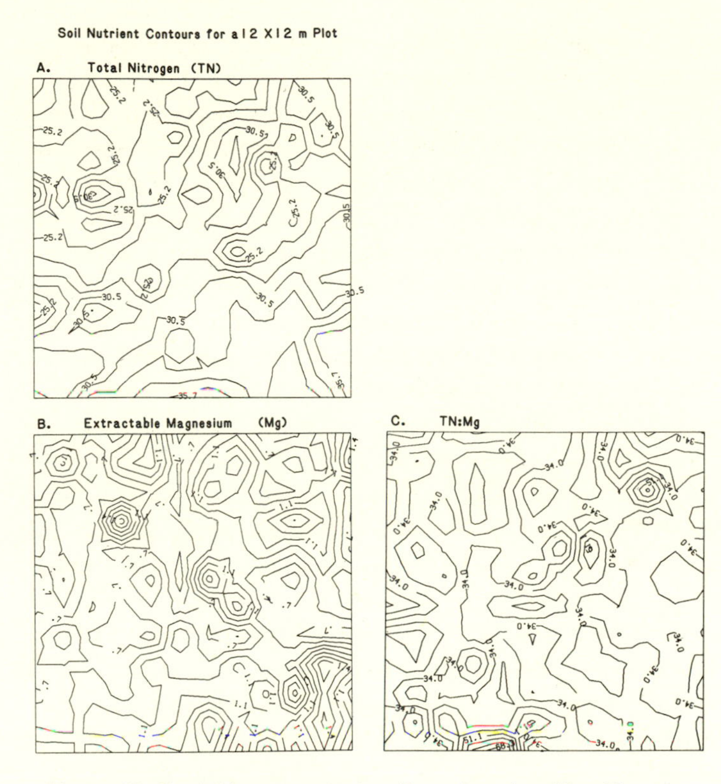

FIGURE 35. Spatial heterogeneity in soil nutrients in a 12 × 12 m plot of sandy soil at Cedar Creek Natural History Area, Minnesota (Tilman, unpublished).

A. Total soil nitrogen, determined with a persulfate digestion technique, varied by 42% within this relatively small area. The contours shown were numerically fitted using a computer contour program. There were 144 sample points.

B. Extractable magnesium varied by over 100% within this same area.

C. The ratio of total nitrogen to extractable magnesium (TN:Mg) also varied by more than 100% within this area. Thus, individual plants within this area could be experiencing vastly different relative rates of supply of N and Mg.

measured because fertilization studies in adjacent sites demonstrated that these were the two most limiting nutrients (McKone, 1980). These maps suggest that plants of the same species living within one or two meters of each other may experience significantly different rates of supply of the limiting nutrients.

Although I know of no maps of nutrients in aquatic habitats which are on an equivalent spatial scale relative to the size of algae, there are numerous physical and biological processes which can cause spatially heterogeneous nutrient distributions in lakes and oceans. Physical mixing processes such as Langmuir circulation cells combined with vertical nutrient stratification could cause spatial heterogeneity on the appropriate scale in aquatic environments. The presence of small-scale heterogeneity implies that a habitat must be characterized both by the mean supply rate of resources and by the spatial variance in the supply rates of resources as experienced by individuals.

The concept of spatial heterogeneity must be carefully considered, for it depends on the spatial and temporal scales used. For instance, different plants in a habitat may have different supply rates of nutrients at any instant, but their average supply rates may not differ. The time scale over which temporal variation is likely to be important is approximately the period between reproductive episodes, because plants may sequester nutrients for future use. Thus, for oak trees, nutrient supply should be averaged over two or three years, whereas for many freshwater algae, the period is about one day. The spatial scale considered should be the area or volume in which an individual obtains resources during the appropriate time period. A small, annual terrestrial plant might have roots covering an area of 0.1 m^2 of topsoil, whereas an oak tree might cover an area of 300 m^2. A planktonic alga might sink through a column of water 1 to 5 m in length and ca. 50 microns in diameter in a day. I call the size of the area in which one individual obtains resources during one reproductive bout its "microhabitat." Given these constraints on spatial and temporal scale, it is possible to make an ecologically meaningful definition of resource spatial heterogeneity: *Spatial heterogeneity is the amount of variance among randomly sampled microhabitats in the supply of resources averaged over the reproductive period of the individual.* These constraints limit the variation considered to variation that is likely to affect the long-term average reproductive success of individuals.

Although much of this chapter deals with the quantitative effects of spatial heterogeneity and resource abundance on interspecific competition, the major qualitative predictions can be visualized from Figure 36. In Figure 36, seven species compete for two essential resources. The positions of the isoclines show that each species is specialized on a different proportion (ratio) of the two resources. A dot marks each of the six two-species equilibrium points, and the lines from each equilibrium point delimit the resource supply regions which lead to each two-species equilibrium. Each equilibrium point is locally stable. As inspection of Figure 36 demonstrates, at most two species can coexist at equilibrium in a homogeneous environment on these two resources. In a spatially heterogeneous environment, both the mean resource supply point and its variance are needed to determine the number of species that can coexist at equilibrium. Let each of the circles (labeled 1-4) in Figure 36 include 99% of the microhabitat to microhabitat resource variation in each habitat, with the mean resource supply point at the center of each circle. (The circle can be thought of as the 0.99 probability contour of a bivariate normal distribution with equal variances and no correlation.) If the regions in which a species can exist do not overlap with this circle, less than 1 in 100 of the microhabitats associated with that habitat are capable of supporting the species, and the species may be considered absent.

Figure 36 reveals two important results. First, for a given degree of spatial heterogeneity, resource enrichment may decrease the number of species that can coexist. For instance, if habitat 1 (Fig. 36) were enriched to the level of habitat 2,

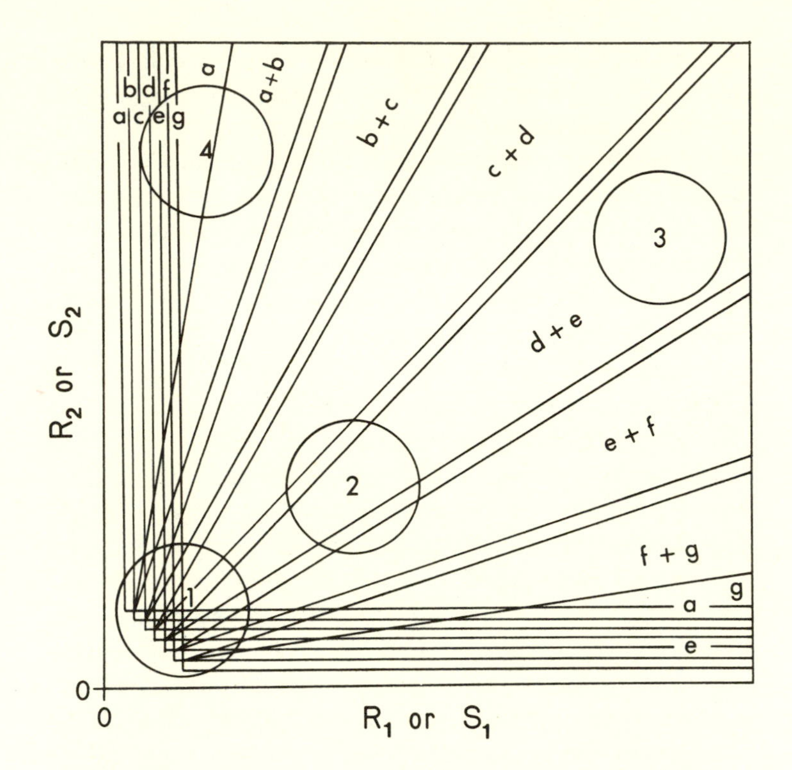

FIGURE 36. Seven species competition for two resources. The ZNGI's for 7 species are labeled *a* to *g*. The six associated regions of coexistence are shown, with "*a* + *b*" meaning that species *a* and *b* coexist in habitats with supply points in that region, etc. Circles represent contours that include 99% of the microsite to microsite spatial variation in resource supply in a particular habitat. Inspection shows that 7 species can coexist in habitat 1, whereas only 2 species can coexist in the more nutrient-rich habitat 3.

The species composition of a habitat is predicted to depend on the relative rates of supply of the resources. For instance, species *d* and *e* should dominate habitats in which R_1 and R_2 have almost equal supply rates, whereas species *f* and *g* should dominate habitats which have a high rate of supply of R_1 but a low rate of supply of R_2.

the number of species would decrease from 7 to 4. Further enrichment, to the level of habitat 3, would decrease this to two coexisting species. Both resources are increased in equal proportion along this enrichment gradient. Species *d* and *e* eventually dominate. The enrichment gradient represented by habitats 1 and 4 (i.e., by enrichment mainly with R_2) leads to dominance by species *a* and *b*. Thus, a second result is that the pattern of enrichment (i.e., the ratio of $R_1 : R_2$) determines which species become dominant.

COMPETITION IN A HETEROGENEOUS HABITAT

In order to explore quantitatively the general phenomena just described, steady-state competition among 40 species competing for two limiting resources in a spatially heterogeneous environment was explicity modeled. The parameters used for these simulations are illustrated in Figure 37A, as is the method of solution. The "kink points" of the ZNGI of the 40 plant species were placed along the line $R_2 = 1 - R_1$, with the kink point of species 1 at (0.75,0.25) and that of species 40 at (0.25, 0.75). The kink points of the ZNGI of the other 38 species were placed in this interval in one of two ways. For the case of "Uniform Spacing," the other 38 species were spaced uniformly between species 1 and species 40. This case is illustrated in Figure 37A, in which the ZNGI of species 1, 10, 20, 21, 30, and 40 are shown. For the case of "Uniform-Random Spacing," the 38 intermediate species were randomly placed within each of 38 intervals of uniform length between the kink points of species 1 and 40.

The positions of the 40 ZNGI's determine the positions of the 39 two-species equilibrium points and also the direction of the consumption vectors associated with each equilibrium point. Each of the 39 two-species equilibrium points is locally stable, because at each equilibrium point each of the two associated species consumes relatively more of the resource that

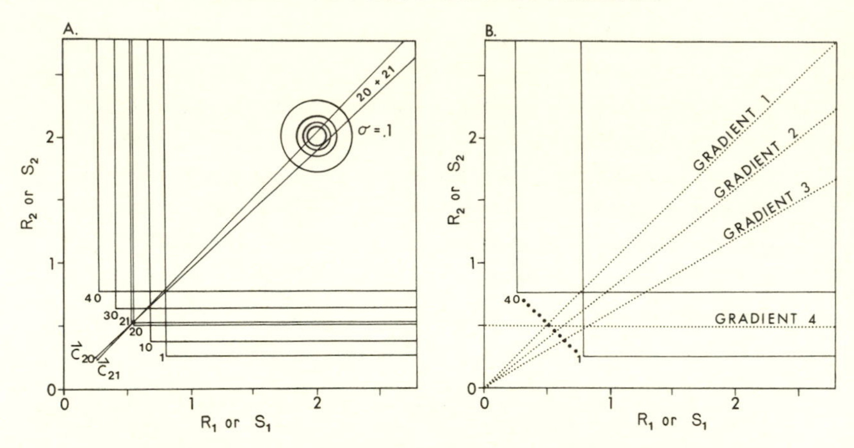

FIGURE 37. A. Competition among 40 species. This figure illustrates the parameters used in the simulations of 40 uniformly spaced species competing for two resources in a spatially heterogeneous environment. The ZNGI's for species 1, 10, 20, 21, 30 and 40 are shown.

B. Enrichment gradients. For cases of competition among 40 species, the effect of resource richness on community structure was explored along four different enrichment gradients, illustrated above.

limits its own growth rate. The region of coexistence for species 20 and 21 is shown in Figure 37A.

The equilibrium densities of all species in a spatially heterogeneous habitat were determined numerically. The procedure used would first generate a bivariate normal distribution with an average resource supply rate of $(\overline{S}_1,\overline{S}_2)$, a correlation coefficient of r, and variances of σ_1 and σ_2 for the S_1 and S_2 components of the spatial variation. This gives a resource supply probability cloud. Probability contours for such a distribution are illustrated in Figure 37A for a case with $\overline{S}_1 = \overline{S}_2 = 2.0$, $r = 0$, and $\sigma_1 = \sigma_2 = 0.1$. The probability distribution so generated was divided into 4000 equally probable regions, each representing resource supply in a particular, spatially homogeneous, microhabitat. The use of 4000 microhabitats to approximate the complete bivariate normal distribution minimized errors that could be introduced into calculations from subsampling. This sample size means that the procedure modeled resource competition in an area 4000 times the size of the microhabitat of the average individual. For each of the 4000 microhabitats chosen, the numerical procedure would determine the equilibrium density of each species in the microhabitat. These equilibrium densities were then summed over all 4000 microhabitats to give the total density of each species in a spatially heterogeneous habitat with an average resource supply of $(\bar{S}_1, \bar{S}_2)$, variances of σ_1 and σ_2, and a correlation coefficient of r.

These data on the population density of each species in the habitat were used to calculate Shannon-Weiner diversity, H', Pielou's evenness index, J (Pielou, 1969), and species richness (for which a species was considered present if its population was at least 0.001 of the total community). The data were also used to generate a dominance-diversity curve (Whittaker, 1965; May, 1975, 1978).

The numerical process described above determines community structure for one particular set of conditions. In order to determine the effect of resource richness on community structure, such calculations are needed for numerous resource supply points along a gradient of resource richness, which I term an enrichment gradient or a resource richness gradient. Because this is an equilibrium model, such gradients summarize both how equilibrium community structure should depend on resource availability and how enrichment should affect community structure. The four gradients used are shown in Figure 37B. For each enrichment gradient, the equilibrium species densities, H', species richness, and J were obtained at each of ca. 80 points along the gradient. For enrichment gradient 1, $\bar{S}_1/\bar{S}_2 = 1$. This represents equal enrichment with both resources. Enrichment gradients 2 and 3 have $\bar{S}_1/\bar{S}_2 = 0.8$ and $\bar{S}_1/\bar{S}_2 = 0.6$, respectively. Enrichment gradient 4 has a slope of 0, with an intercept of 0.5, representing enrichment with only R_1. For enrichment gradients 1-3, the proportional increase in both resources means

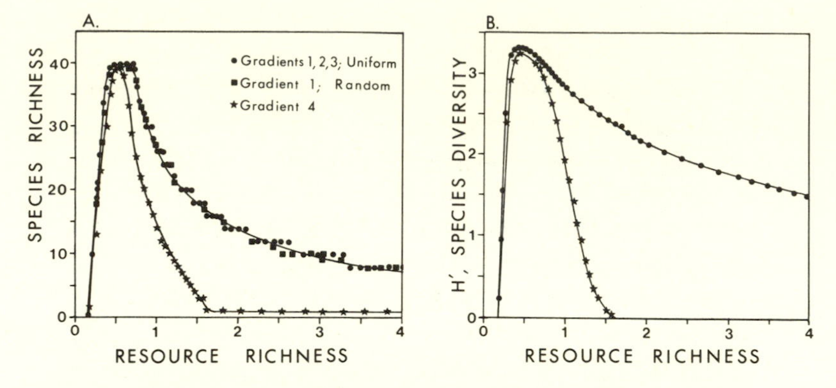

FIGURE 38. A. The predicted dependence of species richness on resource richness for uniformly spaced and randomly spaced communities along enrichment gradients 1 to 4. For all cases, $r = 0$ and $\sigma_1 = \sigma_2 = 0.15$.

B. Shannon-Weiner diversity (H') for these same uniformly spaced communities along enrichment gradients 1-3 (dots) and along gradient 4 (stars).

that about equal numbers of species will be limited by each resource. However, for gradient 4, enrichment by only R_1 means that most species will be limited by R_2. To compare the results of these various enrichment gradients, the level of resource availability is expressed as Resource Richness = $(\bar{S}_1 + \bar{S}_2)/2$.

RESOURCE RICHNESS AND SPECIES DIVERSITY

Figure 38 shows the predicted dependence of species richness and species diversity (H') on resource richness for a case in which $\sigma_1 = \sigma_2 = 0.15$, and $r = 0$ (no correlation between the supply of R_1 and R_2). Species richness rises rapidly with enrichment, reaches a peak species richness in moderately resource-poor habitats, and then declines (Fig. 38A). The resource richness–species richness curve is humped. Species diversity (H') behaves similarly (Fig. 38B). Resource enrichment gradients 1, 2, and 3 (Fig. 37B) give similar results, but enrichment gradient 4 gives a resource richness–species richness curve which falls more rapidly than the curves for gradients 1-3.

This suggests that the diversity of natural plant communities should be maximal in habitats with low nutrient levels and should decrease with increasing productivity. As this is a prediction of a potentially major pattern of diversity in plant communities, it is important to explore its generality. To do this, several simplifying assumptions of the model were made more realistic (Fig. 39). For all modifications, species richness and species diversity (H') gave a humped curve when graphed against resource richness.

Figure 39A demonstrates that a humped curve is obtained for all levels of spatial heterogeneity. For Figure 39A, $\sigma_1 = \sigma_2$, $r = 0$ (i.e., the resource supply probability cloud is circular.) Curves *a-d* show the relationship for heterogeneities (σ) ranging from 0.05 to 0.30. Very heterogeneous habitats lead to a resource richness–species richness curve which is flattened on top (curve *d*, Fig. 39A). Figure 39B illustrates that a humped diversity curve is obtained independent of microhabitat to microhabitat correlation in resource supply. Such correlation in the supply of two resources will cause the resource supply probability cloud to have an elliptical shape. Resource richness–species richness curves for cases with correlation coefficients of 0.66 and 0.88 are shown in Figure 39B. In both cases, and all others tried, neither positive nor negative microhabitat to microhabitat correlation qualitatively changed the shape of the resource richness–species richness curve.

All of the analyses thus far discussed have assumed that the amount of spatial variance is constant, independent of resource richness. However, sampling of almost any natural entity reveals that the variance (σ^2) tends to increase with the mean. Greater spatial variance in more resource-rich habitats will cause the diameter of the resource supply probability cloud to increase with enrichment. Such an increase in diameter with enrichment

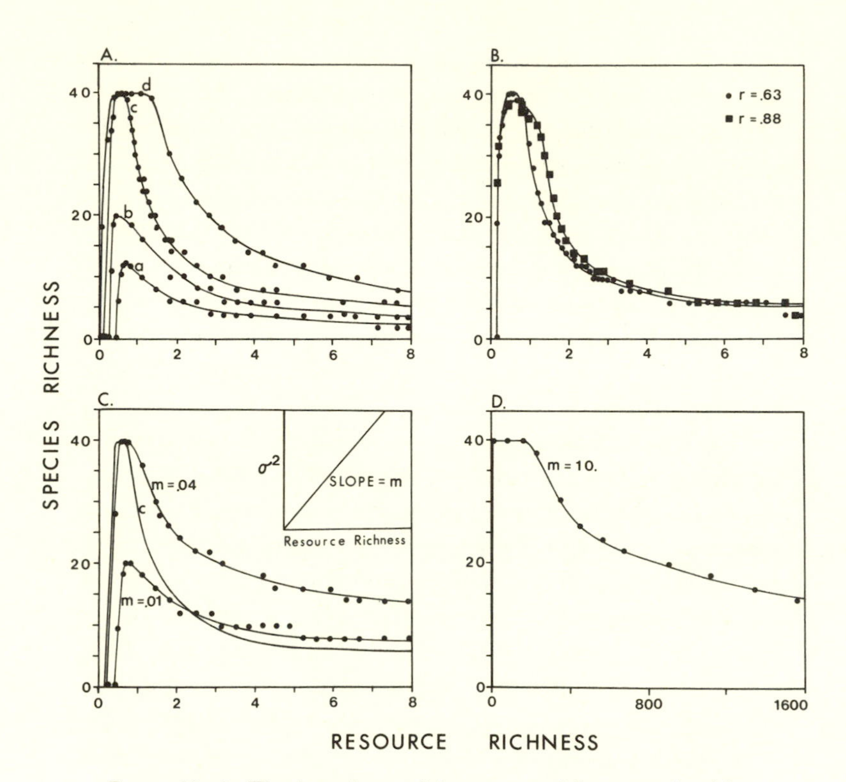

FIGURE 39. A. The dependence of the resource richness–species richness curve on the amount of spatial heterogeneity, where $\sigma_1 = \sigma_2 = \sigma$ and $r = 0.0$, for σ of 0.05, 0.1, 0.15, and 0.3 for curves *a-d*, respectively, all for uniformly spaced communities.

B. Resource richness–species richness curves for $\sigma_1 = \sigma_2 = 0.15$ and $r = 0.63$ or $r = 0.88$, for communities of uniformly spaced species.

C and D. Effect of a linear increase in spatial heterogeneity, σ^2, with resource richness on the shape of the resource richness–species richness curve for a slope, m, of 0.01, 0.04, and 10. Curve c from part A of this figure is shown in part C for comparison. As long as spatial heterogeneity, σ^2, increases no faster than as a linear function of resource richness, this theory predicts a humped resource richness–species richness curve.

might alter the shape of the resource richness–species richness curve.

To explore this possibility, numerous cases (for all, $\sigma_1 = \sigma_2$, $r = 0$) were solved in which the spatial variance, σ^2, was allowed to increase with the average resource richness. As illustrated in the insert in Figure 39C, the spatial variance was assumed to increase in a linear manner with resource richness, with the slope of the relationship being m. This model is comparable to assuming that resources are supplied by discrete, random processes. If $m = 1$, the process would be Poisson. Slopes of 0.01 and 0.04 lead to the humped diversity curves shown in Figure 39C. These may be compared with curve *c* of Figure 39A, which is also shown in Figure 39C. Having spatial variance increase with resource richness broadens the resource richness–species richness curve, but it still has a humped shape. Figure 39D illustrates an extreme example, in which the spatial variance increased 10 times more rapidly than the mean resource availability ($m = 10$). Even with this very rapid increase of spatial variance with the mean availability of the resources, the resource richness–species richness curve is humped. If the variance is allowed to increase at a more rapid than linear rate with resource richness, such as σ^2 proportional to $(\overline{S}_1 + \overline{S}_2)^2$, the resource richness–species richness curve could be flat. Because variance in resource supply is truncated by the impossibility of negative resource levels and by saturating processes at high levels, it seems unlikely that spatial variance would ever increase so rapidly as to give a flat species richness curve. This possibility, though, can easily be tested in the field by looking at the relationships between microhabitat to microhabitat variance, average resource supply rate, and species richness.

The analyses of Figures 38 and 39 assume that there are only two limiting resources in the community. The difference between enrichment gradients 1, 2, and 3 (for which both resources were increased) and enrichment gradient 4 (for which only R_1 was increased) can give some insight into the generality of these results for situations in which there are more than two limiting resources. Within the context of the model of Figure 37, enrichment with both of the limiting resources means that the biomass of the total community will increase but that both of these resources will remain limiting. When this occurs, the effect of enrichment on species richness is as shown in Figure 38A for gradients 1, 2, and 3. In contrast, enrichment with only R_1 means that an increasingly great proportion of the species will be limited by R_2. As this occurs, the community will become increasingly dominated by the one species which is the best competitor for R_2. This leads to a rapid decrease in diversity with enrichment, as shown in Figure 38A for gradient 4. If a community has several limiting resources, enrichment with all but one of the resources would be comparable to enrichment gradient 4 of Figure 38A. The community would be rapidly driven toward dominance by the one species which is the superior competitor for the single limiting resource. However, if only one or a few of the numerous limiting resources were supplied, species richness would fall less rapidly with enrichment beyond the diversity peak than is shown for gradient 4 in Figure 38A.

Figure 40 illustrates this for a community with four limiting resources which starts with all resources being supplied at the rate which leads to maximal species richness. Enrichment with a single resource (such as R_1) leads to decreased diversity as shown. Enrichment with two resources leads to a more rapid decline in species richness, and enrichment with three resources leads to an even faster decline in species richness, comparable to gradient 4 of Figure 38A. The least rapid decline in diversity occurs when all four resources are added. The curve for addition of all resources is comparable to that of gradients 1, 2, and 3 of Figure 38A for which both limiting resources were added. This curve, which has the most gentle decline in diversity with enrichment of all the cases in Figure 40, is probably the least likely ecologically. Enrichment of any natural plant community

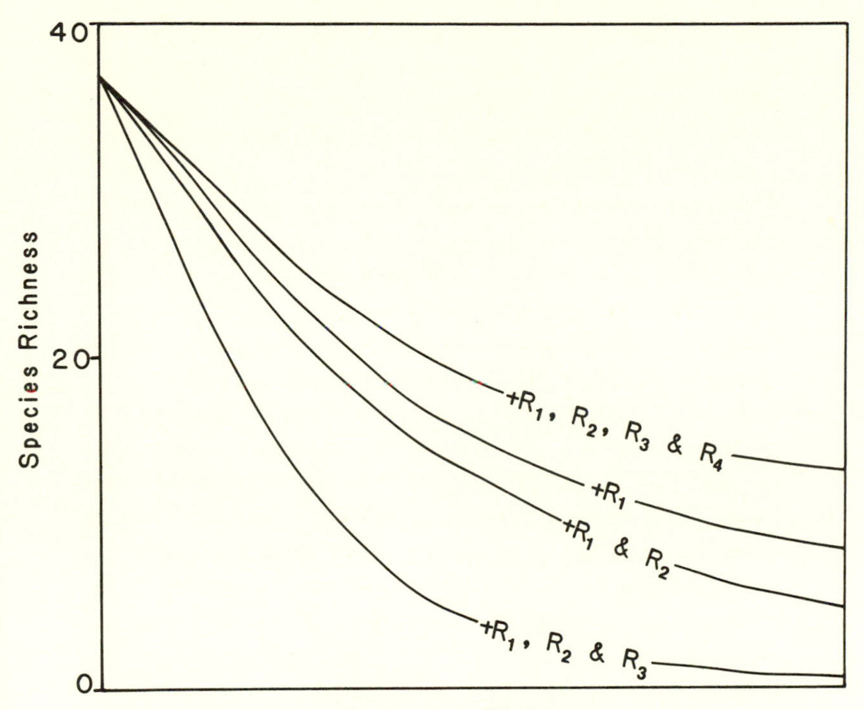

FIGURE 40. The hypothesized effect of resource richness on species richness for a community with four limiting resources. Patterns of enrichment that drive a community toward an "edge" (for which all species will be limited by the same resource) should lead to more rapid declines in diversity than enrichments for which all resources are increased by the same proportion (for which there will still be several different limiting resources).

with nutrients should always lead to a decline in diversity which is more rapid than this because one resource—light—is never added. As biomass increases with fertilization or with natural increases in nutrient availability, light becomes increasingly limiting. Thus, fertilization with all limiting nutrient resources should lead to a community which is eventually dominated by the species which is the superior competitor for the non-nutrient resource, light. Even if the spatial variance in the supply of all nutrient resources were to increase more rapidly than the square of their mean, there should still be a humped relationship between resource richness and species diversity because light would become the only limiting resource. This further suggests that a humped resource richness–species richness curve should be generally expected in plant communities.

In total, these theoretical results suggest that a humped resource richness–species richness curve may be a general and robust prediction of an equilibrium theory of plant competition for resources. Figure 36 illustrates the mechanism leading to the humped curve. Because nutrients are required for plant growth, extremely nutrient-poor habitats will have only a few, sparse species—those which are capable of surviving under such very low nutrient conditions. Slightly more nutrient-rich habitats will allow more species to survive. Thus, the ascending side of the resource richness–species richness curve is caused by the increased number of species that can live in the more resource-rich habitats. The maximal number that can coexist depends on the level of spatial heterogeneity, i.e., on the variance in resource availabilities between microhabitats. Peak diversity occurs at a mean resource supply rate about equal to the mean resource requirement for survival of each species in the habitat, because it is at this point that the resource supply probability cloud overlaps the greatest number of regions of existence and coexistence of the species present. The decline in species diversity for habitats with a resource richness greater than at the diversity peak is caused by interspecific competition. In the absence of other species, each species is able to survive in all such habitats. However, each species is a superior competitor for only a small range of resource supply ratios. Enrichment reduces the effective range of resource ratios, and increasingly fewer species have their region of coexistence included within the probability cloud of resource supply points. This occurs because the width of the region of coexistence for any two species increases linearly with resource richness (see Fig. 36), but the width of the probability cloud of resource supply points may be constant or may increase, but probably no faster than the square root of resource richness (because the width of this probability cloud is proportional to σ, not σ^2). If spatial heterogeneity of nutrient supply were to increase more rapidly than in proportion to the square root of resource richness, a humped diversity curve could still occur if plants became light limited, giving a resource richness–species richness curve comparable to gradient 4 of Figure 38A.

RESOURCE RICHNESS AND COMMUNITY STRUCTURE

The analyses just presented include additional information on the structure of plant communities. One measure of the relative abundance of species in a community is Pielou's evenness or equitability index, J. For all the analyses performed, graphs of J against resource richness give downward tending curves. This means that both species richness and equitability decline with enrichment according to this equilibrium model.

The effect of resource richness or resource enrichment on community structure is more fully illustrated using dominance-diversity curves (Whittaker, 1965; May, 1975) or Preston-type log abundance curves (Preston, 1948). For brevity, only dominance-diversity curves will be shown. Figure 41 presents four sets of dominance-diversity curves (A-D). Each set shows the dominance-diversity relationship at four different points along a resource enrichment gradient. Part 1 of each set is for a community that is on the ascending portion of the resource richness–species richness curve, part 2 is for a community at the diversity peak, and parts 3 and 4 show communities on the descending portion of the curve. Set A of Figure 41 gives dominance-diversity relationships for communities of uniformly spaced species on enrichment gradient 1, with $\sigma_1 = \sigma_2 = 0.15$ and $r = 0$. This gives a resource supply probability cloud that is circular, similar to that depicted in Figure 37A. Set B of

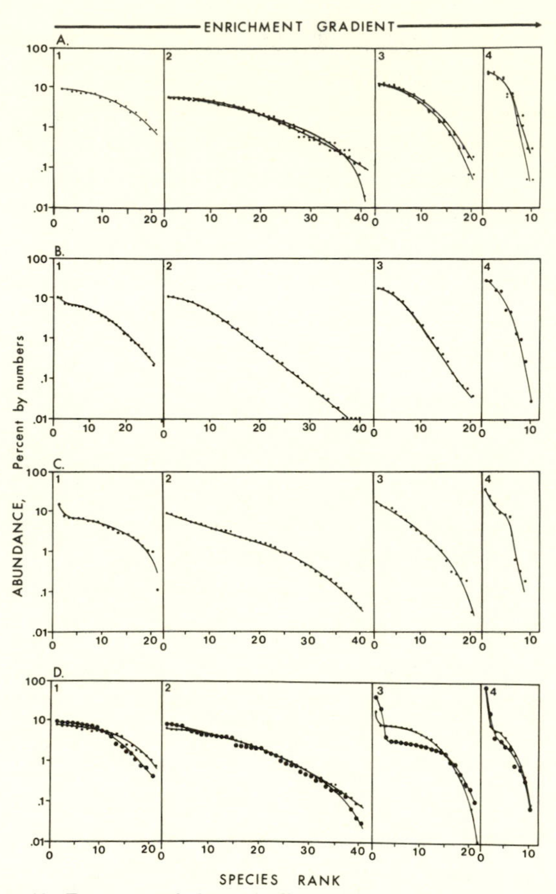

FIGURE 41. Four sets of theoretically predicted dominance-diversity curves are shown. Set A is for species with uniformly spaced resource isoclines, with $\sigma_1 = \sigma_2 = 0.15$ and no correlation in resource supply ($r = 0$). The four parts of set A show four different points along enrichment gradient 1, with point 1 being a very low resource richness habitat, below the diversity peak; point 2 being at the diversity peak; and points 3 and 4 being for resource richnesses beyond the diversity peak. Set B shows curves at comparable points along an enrichment gradient but with strong positive correlation between the supply of the two resources ($r = 0.88$). Set C is comparable to set A except that the resource isoclines were uniform-randomly spaced and the length of the consumption vector was random. Set D shows two cases with enrichment gradient 4. Small dot curves are otherwise identical to set A and large dot curves to set C.

Figure 41 shows dominance-diversity curves for a similar case, except that $r = 0.88$, giving an elliptical shape to the resource supply probability cloud. Set C illustrates dominance-diversity curves for a case similar to that of set A, except that the species were uniform-randomly spaced and the length (but not the slope) of the consumption vector was a random variable. Set D shows two cases for which enrichment gradient 4 (enrichment with only R_1) was used. The curves of set D that are drawn through small dots are for a case otherwise identical to set A, and the curves through large dots are otherwise identical to set C.

These results illustrate several important points about community structure as predicted by an equilibrium model of resource competition. First, unless random elements are added to the model, the model does not seem to predict previously hypothesized distributions (log-normal, broken stick, or geometric; Whittaker, 1965). None of the dominance-diversity curves of Figure 41A and B have the inflection point necessary for a log-normal or broken stick distribution. The distributions tend to be approximately geometric, but are often convex with respect to the origin. When random factors are added (sets C and D), the communities at the diversity peak are not greatly affected, but the communities beyond the peak tend to have an inflection point. This is especially true for enrichment gradient 4 (set D), for which most species are limited by the same resource, thus favoring the one species which is the best competitor for that resource. The tendency for random elements to add an inflection point to a basically geometric relationship (compare curve 4 for sets A and C) is consistent with May (1975).

SUMMARY OF RESULTS OF THEORY

These results demonstrate that niche diversification is theoretically possible for communities of plants competing for just two essential nutrients. Contrary to the generalizations of Grubb (1977), Connell (1978), and Huston (1979), the similarity of resource requirements of plants does not limit, in theory, the species diversity of equilibrium communities. The theory of resource competition in a heterogeneous environment makes several predictions, which I here summarize.

(1) The resource richness–species richness curve should have a hump, with highest diversity occurring in relatively resource-poor habitats.
(2) The ascending portion of the resource richness–species richness curve should be steeper than the descending portion.
(3) Communities with resource levels near the diversity peak should have many relatively co-dominant species, whereas more resource-rich, lower diversity communities should be dominated by a few species, with most species being rare. This is comparable to saying that evenness (*J*) should be highest near the diversity peak and decrease with resource richness.
(4) For a given level of resource richness, increased spatial heterogeneity should lead to increased species richness, with the most marked effects in resource-poor habitats.
(5) The ratio of resource supply rates should determine which species are dominant in a community.

This last point may need illustration. Enrichment gradients 1 to 4 of Figure 37B represent habitats with ratios of resource supply (S_1/S_2) of 1, 0.8, 0.6, and "0" where the supply ratio of gradient 4 approaches zero as increasingly rich habitats are considered. Species 20 and 21 tend to dominate gradient 1 (for $S_1/S_2 = 1.0$), whereas species 17 and 18 tend to dominate for $S_1/S_2 = 0.8$, species 13 and 14 tend to dominate for $S_1/S_2 = 0.6$, and species 1 tends to dominate gradient 4. The ratio of the supply rates of the limiting nutrients determines which species are dominant. The extent of their dominance is determined by the amount of spatial heterogeneity (σ) and by the resource richness of the habitat. The more resource rich a habitat, the greater will be the dominance of the species favored by a certain nutrient ratio. This can be easily visualized from Figure 36 in which the position of the resource supply probability cloud (represented by the 0.99 contour of the distribution) is seen to determine the species composition of the community.

Thus, a theory of resource competition in a heterogeneous environment has the potential to predict not only the dependence of community structure on resource richness, but also which species should be dominant in a habitat, and which species should become dominant with any given pattern of enrichment. As such, it offers many opportunities for its possible falsification.

FIELD OBSERVATIONS AND EXPERIMENTS

There are three types of information on natural communities which may be relevant to this theory: (1) correlations between community diversity and resource (nutrient) availability; (2) observed effects of inadvertent nutrient enrichment (eutrophication) on community diversity; and (3) observed effects of nutrient enrichment on community diversity in controlled experiments. I will consider them in this order.

I know of three studies which allow comparison of the species diversity of different habitats within a geographic region with the level of resource availability for each habitat (Fig. 42). Figure 42A shows the relationship observed by Beadle (1966) between HCl-soluble soil phosphorus and the total number of genera of Australian xeromorphic and rainforest genera. The curve drawn is my (admittedly optimistic) eyeball fit to the data. Although genera richness is not species richness, Ashton (1977) reported a constant species to genus ratio for numerous tropical lowland forests. If this is so for the Australian flora,

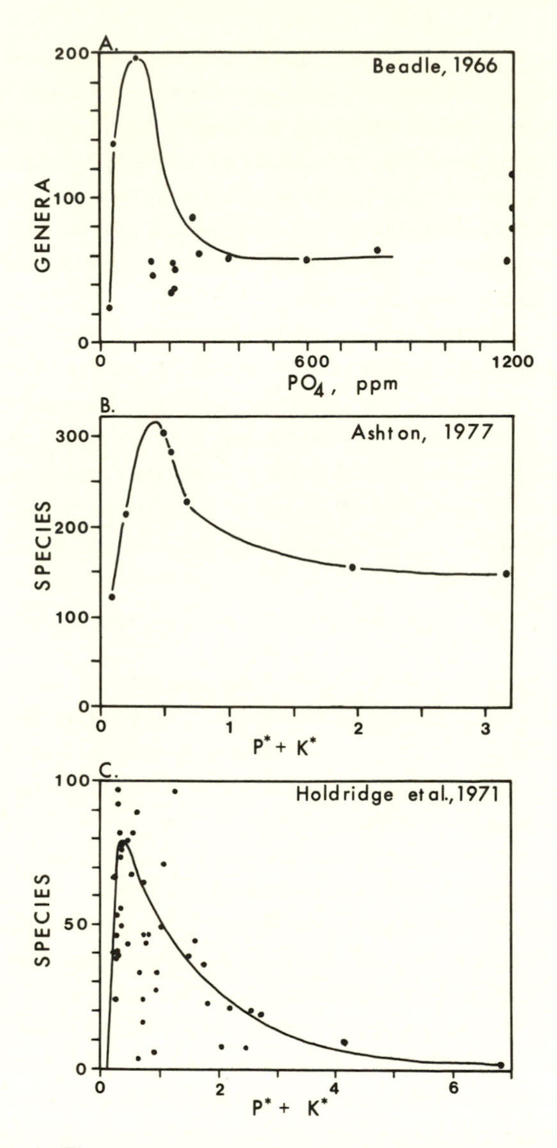

FIGURE 42. The observed relationship of species richness (or genera richness for part A) to resource richness for three natural plant communities. The curves are hand drawn to data, not fitted.

Beadle's results are suggestive of a humped species richness–resource richness curve. The only divergence occurs at such high levels of phosphate that phosphate may no longer be a limiting nutrient. Figure 42B shows a similar relationship, this from data in Ashton (1977). The number of species of Malaysian rainforest woody plants shows a humped curve. To analyze Ashton's data, the observed potassium concentration in any sample was divided by the average over all samples. Phosphorus concentrations were similarly normalized. Then, the normalized phosphorus (P^*) and potassium (K^*) were summed to provide an index of resource richness. Ashton's data are strikingly similar to theoretical predictions. The same index of resource richness was used to analyze the data collected by Holdridge *et al.* (1971) for Costa Rican forests (Fig. 42C). Although there is scatter, the 46 data points show a strong tendency for communities with high resource richness to contain only a few species, and for the most diverse communities to be associated with low nutrient levels. However, the data do not show a definite ascending portion to the resource richness–species richness relationship. The data of Holdridge *et al.* (1971) have been more thoroughly analyzed by Huston (1980), whose analysis shows a clear tendency for diversity to decrease with resource richness.

Other correlational evidence is also relevant to this analysis. Two of the most species-rich plant communities in the world occur on very nutrient-poor soils (the Fynbos of South Africa and the heath scrublands of Australia). Nearby communities on more nutrient-rich soils have a much lower plant species richness (Goldblatt, 1978; Kruger and Taylor, 1979; Specht and Rayson, 1957; Specht, 1963). Bond (1981) reported a humped relationship between species richness and standing crop for the Fynbos of South Africa. Similarly, Al-Mufti *et al.* (1977) reported a humped relationship similar to that predicted in this paper between species diversity in terrestrial plant communities and the total standing crop (including litter). If standing crop is indicative of soil nutrient status or productivity, these data would support this theory. Additionally, Grime (1973) reported maximal plant diversity on pasture soil of intermediate fertility, with both very low fertility and very high fertility soils having lower species richness. Young (1934) noted that, in undisturbed forests of the Adirondack Mountains, richer soils supported a less diverse flora. A similar pattern was reported by Mellinger and McNaughton (1975) for old field communities.

If water is a limiting resource on prairie or is correlated with the availability of limiting resources such as soil nutrients, the work of Dix and Smeins (1967) on a North Dakota prairie shows a humped diversity curve which also has highest evenness (J) at the species richness peak. A similar pattern is reported by Whittaker and Niering (1975).

There are numerous aquatic examples of the effects of resource richness on community structure. The nutrient-poor, unproductive waters of the Sargasso Sea support a very diverse phytoplankton assemblage, whereas more productive and nutrient-rich temperate oceans support a much less diverse flora (Steeman-Nielsen, 1954; Fischer, 1960; Guillard and Kilham, 1977). Nutrient-rich, marine upwelling regions are relatively species-poor compared to nearby, non-upwelling regions (Dugdale, 1972; Smayda, 1975; Blasco, 1971; Nelson and Goering, 1978). Similar relationships have been reported for animal communities. For instance, Whiteside and Harmsworth (1967) observed that species diversity (H') of chydorid cladocera decreased with increasing algal productivity in both Danish and Indiana lakes.

The diversity difference between temperate and tropical waters is made even more interesting by Hulburt's (1982) studies of Casco Bay (Maine) and the Sargasso Sea. Casco Bay, in which nitrogen is the only limiting nutrient for two to three summer months, is an almost pure monoculture of one species, the diatom *Skeletonema costatum*. It makes up from 95% to over

99% of the community during this time. In contrast, nitrogen, phosphate and, at times, silicate are limiting in the Sargasso Sea. Hulburt reports 5 to 7 species co-dominating this habitat, with many other rarer species co-occurring with them. The more resource-rich habitat in which there is only one limiting resource is essentially a monoculture, even though there is surely some spatial and temporal variability in nitrogen levels. The more nutrient-poor habitat, in which there are several limiting resources, has numerous coexisting species. These observations are consistent with the theory of resource competition in a spatially structured habitat developed in this chapter.

The cultural eutrophication of lakes, rivers, estuaries, and coastal marine regions has had two consistent effects: an increase in the rate of primary productivity and a decrease in the species diversity of the phytoplankton. The data reported by Williams (1964) suggest that eutrophic waters have both a lower diversity and a lower evenness (*J*) than associated undisturbed waters. The work of Williams (1964), Patrick (1963, 1967), Patten (1962), Schelske and Stoermer (1971), and the review in Russell-Hunter (1970) are but part of the literature documenting the decrease in species richness associated with nutrient additions to algal communities. A decrease in species richness with enrichment is consistent with the theory developed in this chapter because most enrichments are large, and diversity is only predicted to increase with resource richness through a very narrow range, and only in very resource-poor habitats.

DIRECT EXPERIMENTAL EVIDENCE

There have been several studies in which pastures, grasslands, or old-field communities have been fertilized and the results observed for periods ranging from one year to >120 years. I will first summarize the results from the short-term experiments, and then provide a more complete analysis of the results from the Park Grass Experiments of Rothamsted,

England. In an eight-month experiment, the application of N-P-K (6-12-12) fertilizer to an eight-year-old field led to decreases in both diversity and evenness compared to control plots (Bakelaar and Odum, 1978). Watering and fertilization of a short-grass prairie for a three-year period also resulted in a significant decrease in species diversity and evenness (Kirchner, 1977). A 17+ year study of the effect of manuring, cutting, and grazing on the botanical composition of natural pastures (Milton, 1934, 1940, 1947) showed major decreases in species richness with fertilization.

Willis and Yemm (1961) and Willis (1963) studied the effects of nutrients on the vegetation of the sandy coastal dunes at Braunton Burrows, England. In experiments on turf transplanted from damp pasture, unfertilized transplants (watered with deionized water) averaged 22 species at the start and 21 species at the end of the experiment 39 weeks later, whereas transplants watered with a complete nutrient solution (containing N, P, K, Mg, Ca) decreased from a species richness of 21 to a richness of 5. Similarly, those watered with only N, P, and K decreased from 21 to 5 species. For turf transplanted from dry pasture, unfertilized controls averaged 16 species at the end of the experiment, whereas those fertilized with complete nutrients or with N, P, and K averaged 6 and 2 species, respectively. Experiments performed with undisturbed natural vegetation yielded similar, but less dramatic, results (Willis, 1963). Species richness declined from an average of 23 to an average of 17 for fertilized plots, whereas unfertilized plots maintained their original species richness of 24 throughout the experiments.

The Park Grass Experiment

The most long-term ecological experiment with which I am familiar is the Park Grass Experiment at Rothamsted, England, which was initiated in 1856 and which continues to this day.

An eight-acre pasture that had been used for grazing land for the previous ca. 200 years was divided into 20 plots, two serving as controls, the others receiving a particular fertilizer treatment once each year (Lawes and Gilbert, 1880). Although started as an agricultural experiment to determine the effects of various types of fertilization on the yield of hay, Lawes and Gilbert (1880) observed such dramatic changes in species composition during the first few years that they concluded that the experiments were of greater interest to the "botanist, vegetable physiologist, and the chemist than to the farmer." In 1862, Lawes, Gilbert, and Masters (1882) made their first quantitative survey of the species composition of the plots, and these botanical surveys were repeated in 1867, 1872, 1877 (Lawes, Gilbert, and Masters 1882), and in 1903, 1914, 1919, 1920, 1926, 1936, 1948, and 1949 (Brenchley and Warington, 1958), although some plots were not surveyed in each of these later years. The qualitative trends observed in these data have been discussed in several papers (Lawes and Gilbert, 1880, 1900; Lawes, Gilbert, and Masters, 1882; Brenchley, 1924; Brenchley and Warington, 1958; Thurston, 1969; Silvertown, 1980), but this rich data set has not been thoroughly analyzed by ecologists (see, however, Chapter 6).

Figure 43 shows the observed trends in Shannon-Weiner diversity and evenness in the unfertilized plots and in the plots receiving complete fertilization for the period from 1862 to 1949. The unfertilized plots reveal no significant trend in either diversity or evenness. With complete fertilization, there was a very significant decline both in diversity and in evenness. Subplots that were limed in 1903 to overcome pH change associated with long-term ammonia fertilization also showed significant declines in diversity and evenness. Another plot received ammonia fertilization until 1897, during which time there was a significant decline in species diversity and evenness. The declines in diversity and evenness stopped when fertilization was terminated.

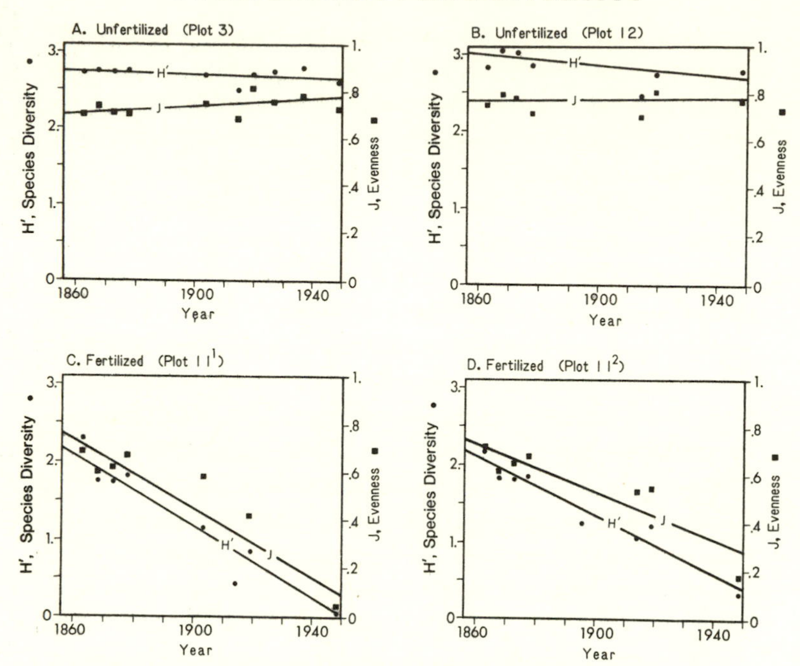

FIGURE 43. A and B. Diversity (H′) and evenness or equitability (J) in two unfertilized Rothamsted plots. Note the lack of major changes throughout ca. 100 years.

C and D. Diversity and evenness in plots receiving complete mineral fertilizer. There was a significant decrease in both diversity and evenness in these plots. Plot 11^1 received ammonium sulfate, superphosphate, potassium sulfate, sodium sulfate, and magnesium sulfate. Plot 11^2 received these and also sodium silicate.

Similar decreases in species diversity were observed in other fertilized plots at Rothamsted. However, theory does not predict that all fertilizations should lead to decreased species diversity. Only the addition of one or more nutrients which are limiting the growth of at least some species should result in changes in community structure. A good measure of the extent to which a given community is limited by a particular nutrient should be the increase in productivity with fertilization by the nutrient. Assuming that the soil of the Rothamsted plots was sufficiently nutrient-rich before fertilization to be beyond the diversity peak, there should be a negative relationship between productivity and diversity for these plots. The assumption that the plots would have all started beyond the diversity peak seems justified by the relatively high biomass (ca. 300 grams of dry plant tissue per m^2) and the complete ground cover observed in the unfertilized plots.

The relationships between species richness and productivity (estimated as the biomass harvested in the first of two annual cuts) for all the Rothamsted plots for several years from 1862 to 1949 are shown in Figures 44 and 45. These relationships were independently analyzed by Silvertown (1980). For each year, two sets of curves are shown. The thin line through stars gives the dependence of species richness on productivity for plots fertilized with ammonia (which tends to make the soil more acidic because plants maintain charge balance by exchanging H^+ ions for ammonia when ammonia is taken up). Dots and thick lines show the relationship for all other treatments, including no fertilization. In all cases, there is a negative relationship, indicating that species richness declines with increases in productivity. Comparison of the slopes of these curves reveals no significant change in this relationship from 1862 to 1949, as well as no significant difference in the slope when comparing plots which were acidified with ammonia with those not acidified. This striking result suggests that the mechanism leading to decreased species richness with increased productivity is sufficiently general that it is unaffected by year-to-year variations in rainfall, mean annual temperature, and plot-to-plot variations in the relative abundances of various limiting nutrients. Species richness seems to be a decreasing, fairly linear function of productivity. As already discussed, productivity is a parameter which represents the integrated effects of resource limitation on plant communities, and can be considered an index of resource richness. Thus, the results of the Rothamsted experiments suggest that a given increase in

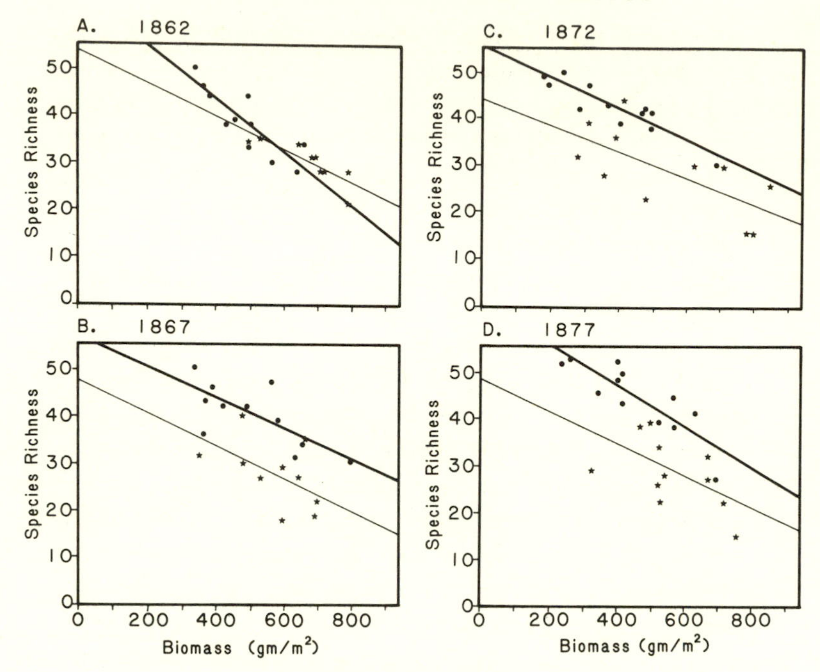

FIGURE 44. The observed dependence of species richness on biomass for the Rothamsted plots in 1862, 1867, 1872, and 1877. Stars are for the plots which were fertilized with nitrogen as ammonia. The thin line running through these points is a regression line. Dots and the thick line show the data and regression for the plots which either were not fertilized with nitrogen or which received nitrogen as nitrate. In all cases, species richness decreases with resource richness, with the slope of the relationship being independent of the type of nitrogen applied. The shift in the intercept of the regression line is probably caused by pH changes associated with ammonia fertilization. For a more complete statistical analysis of these data, see Silvertown (1980).

resource richness has led to the same decrease in species richness throughout the last 100 years. These results further suggest that soil pH has a simple, additive effect on the relationship between species richness and productivity, because the acidified plots differed only in the intercept of the regression line, not in the slope. This difference in intercepts may reflect the number of species which can survive in the habitats in the absence of inter-

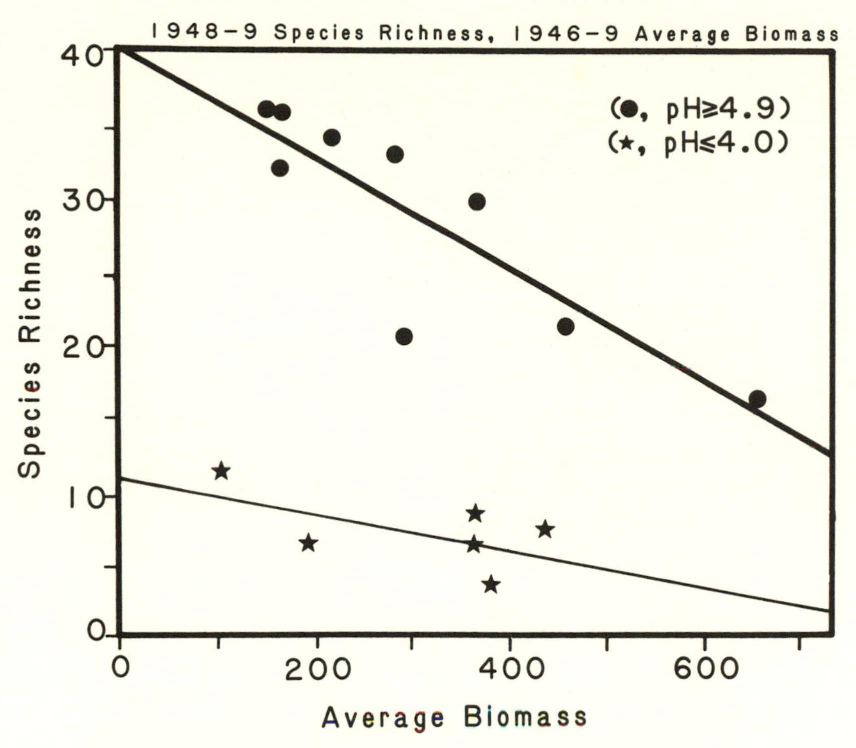

FIGURE 45. The dependence of 1948 and 1949 species richness on the 1946 to 1949 average biomass of Rothamsted plots for which pH measurements are available. All the plots with a pH less than 4 received nitrogen as ammonia, whereas none of the plots receiving nitrogen as nitrate had a pH less than 4.9. For both data sets, species richness tends to decline with increases in plot biomass. The slopes of these regressions do not differ significantly.

specific competition. The increasingly acidic nature of the plots receiving ammonia fertilization probably meant that fewer species were physiologically capable of surviving in those plots. But among those species which could survive, changes in resource richness had the same effect on species richness as in the non-acidified plots. These results suggest that a single mechanism led to decreased diversity in these plots, independent of soil pH and other physical factors. The decrease in species richness with fertilization and the simplicity of the pattern are consistent with the theory of resource competition in spatially heterogeneous habitats which was presented in the preceding pages of this chapter.

From 1862 to 1949, there were 278 plant surveys made in the various fertilized, limed, unlimed, and control plots at Rothamsted. For each of these data sets, evenness, J, was calculated. A graph of evenness against species richness is shown in Figure 46, along with the curve theoretically predicted for enrichment gradient 4 of Figure 37B. This illustrates that the resource-rich (and thus species-poor) communities at Rothamsted have low evenness, whereas the resource-poor communities are more equally dominated by numerous species. This result is qualitatively consistent with the predictions of the theory developed in this book.

A further analysis of Rothamsted community structure may be provided by dominance-diversity graphs for the experimental plots. May (1978) showed five dominance-diversity curves for the fertilized Rothamsted Plot 11^1. Figure 47 shows dominance-diversity graphs for unfertilized Rothamsted Plot 12 and fertilized Plot 11^2. The dominance-diversity curves for the unfertilized plot are approximately geometric with a tendency toward the convex geometric distribution predicted by resource-competition theory for communities near the diversity peak. (Compare Fig. 47A with sections 2 of Fig. 41.) Figure 47B shows all the data reported for Plot 11^2. These show approximately convex-geometric distributions which become increasingly geometric with long-term fertilization. Thus, the dominance-diversity patterns exhibited in the Rothamsted experiments are generally consistent with the theoretical predictions of Figure 41. Although inconsistent patterns could be used to refute the model, the apparent agreement between theory and observation is only weak support for the model because alternative models are likely to make similar predictions.

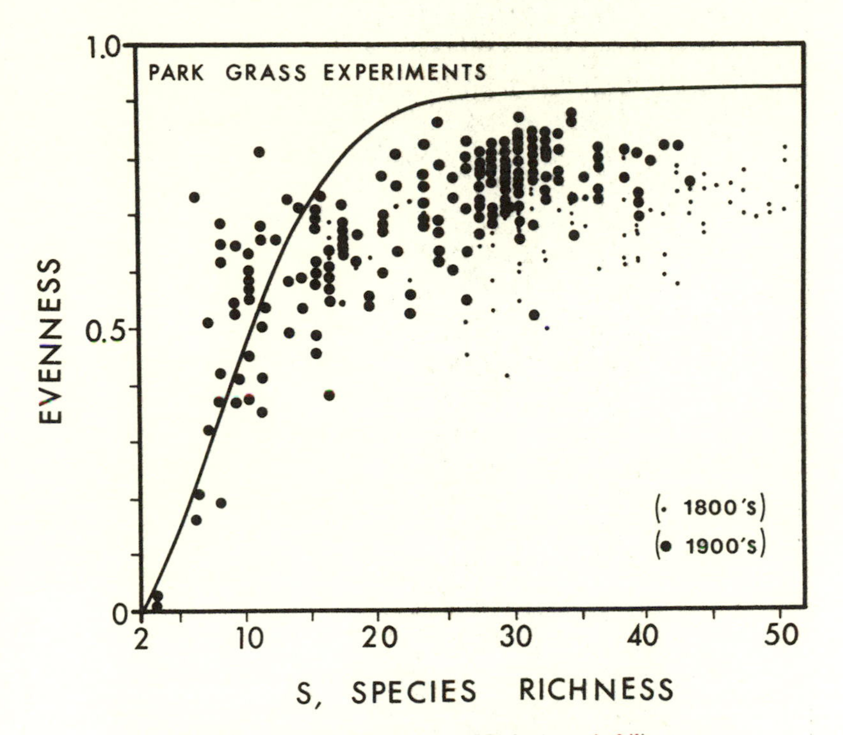

FIGURE 46. The observed dependence of Pielou's equitability or evenness index, J, on species richness, for all the Rothamsted plots over all the years of the experiments. There are 278 data points. There is a tendency for the most heavily fertilized plots, which have the lowest species richness, to have low evenness. This is consistent, in a qualitative way, with the predictions of the theory developed in this chapter. However, such a dependence of evenness on species richness may also be predicted by other models. The curve shown is *not* fitted to the data, but is theoretically predicted by the model of resource competition for enrichment gradient 4 with equal variances and no correlation in the supply of the two resources. Non-zero correlation coefficients lead to different asymptotic evennesses.

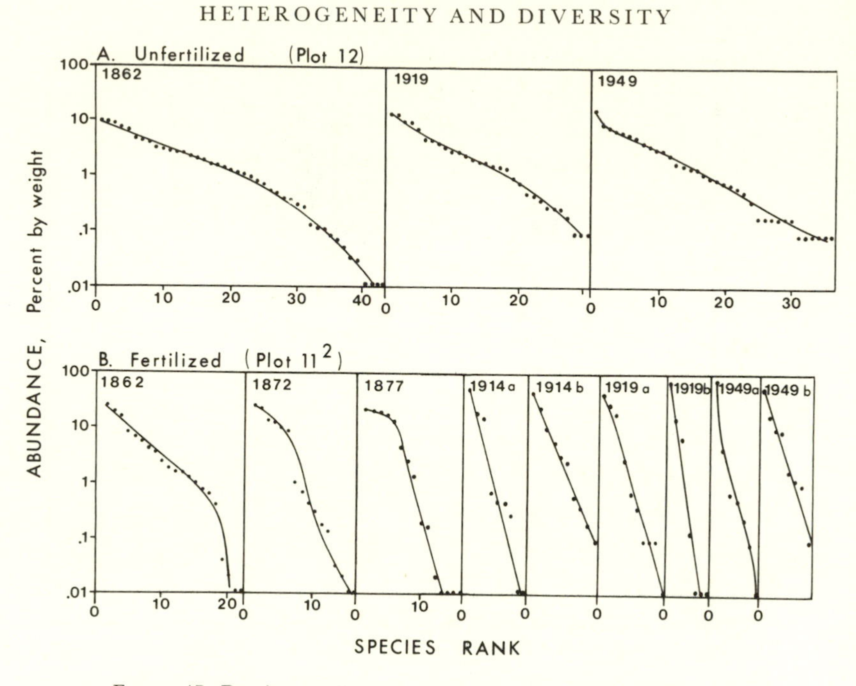

FIGURE 47. Dominance-diversity curves for some of the Rothamsted plots. Part A shows curves for unfertilized plots and part B shows curves for fertilized plots. The curves of part B which are labeled *a* are unlimed and those labeled *b* are from limed plots.

In total, the Rothamsted experiments show many trends which are consistent with the predictions of an equilibrium model of resource competition. The dramatic decreases in species richness and evenness (J) with enrichment are substantial experimental evidence in agreement with the model. Changes in species composition in response to different patterns of enrichment, which are discussed in the next chapter, are also consistent with theory.

IMPLICATIONS AND EXTENSIONS

In tropical and temperate forests, grasslands, lakes, rivers, and oceans, high diversity communities are associated with nutrient-poor habitats. Species-poor communities are most often found in either extremely nutrient-poor or nutrient-rich habitats. The addition of nutrients to plant communities has consistently led to decreased species diversity, with the pattern of enrichment (i.e., the ratio of nutrients supplied) influencing which species become dominant, as is discussed in Chapter 6. These observations, summarized in the preceding section, are consistent with the predictions of an equilibrium theory of resource competition in spatially heterogeneous environments.

The analysis presented in this chapter can be considered an extension of the niche diversification hypothesis. It demonstrates that it is possible for numerous species to coexist, at equilibrium, in a spatially heterogeneous environment if the species differ in the proportion of nutrients that they require. The required proportion is reflected in the position of a species' growth isocline and the slope of its consumption vector. The slope of the consumption vector gives the long-term average proportion (or ratio) of nutrients consumed, and thus should approximate the ratio of nutrient concentrations contained in the plant. Woodwell, Whittaker, and Houghton (1975) and Garten (1978) have documented patterns of species separations in relation to nutrients that are consistent with these predictions.

Two predictions made by the theory presented in this chapter are sufficiently robust that they will not be qualitatively changed by relaxation of many of the simplifying assumptions inherent in the model:

(1) The resource richness–species richness curve should be humped, with the ascending portion steeper than the descending portion.
(2) The ratio of resource supply rates should strongly influence which species are dominant in a community.

This assertion is supported by the following. First, it might be argued that the two predictions made above depend on having an isocline with a perfect right-angle corner. Although allowing the growth isoclines to have rounded corners instead of right-angle corners will move equilibrium points slightly, this will not affect their stability or the general shape or position of the region of coexistence associated with each equilibrium point, and thus will not qualitatively affect the predictions listed above. Figure 39A and B demonstrates that neither the level nor pattern of spatial heterogeneity changes the predictions listed above; nor are these predictions affected by the pattern of placement of these species along a resource gradient (compare uniform and random spaced cases in Fig. 38A). A humped resource richness–species diversity curve is obtained for all cases in which spatial variance (σ^2) increases no faster than as a linear function of resource richness. And, as already discussed, a humped curve could be obtained even if variance were to increase faster than linearly as long as there is at least one resource which becomes more limiting with enrichment. However, these predictions may depend on two assumptions inherent in the equilibrium approach used. The model assumes that species consume resources in the proportion required for balanced growth. If a species were to consume excess quantities of a resource which did not limit its growth (i.e., if a species were to hoard a non-limiting resource), the region of coexistence would become smaller and the equilibrium point could become unstable, with the outcome of competition depending on the past history of each microhabitat. (See Figs. 25 and 26.) Secondly, the model assumes that mortality rate does not depend on resource richness. If increased resource richness led to increased mortality rates for all species, relaxation of this assumption would probably have little effect. However, if increased resource richness led to changing patterns of mortality, the second prediction made above would be less likely to hold.

This approach may be distinguished from others that have been used to explore the "paradox of enrichment." Rosenzweig

(1971) suggested that enrichment may decrease species diversity because enrichment would destabilize consumer-resource interactions. In the model presented in this chapter, enrichment has no effect on the local stability of equilibrium points. The level of resource richness determines only how many locally stable equilibrium points are associated with the resource supply probability cloud. Rosenzweig's analysis applies only to two-species interactions, whereas that presented in this chapter can apply to interactions among any number of species. His analysis assumes that the prey isocline is "humped," an assumption that this analysis does not make. Riebesell (1974) used an approach similar to Rosenzweig (1971) to show that enrichment can lead to the extinction of one of two species competing according to the Lotka-Volterra equations. Huston (1979) used simulations of Lotka-Volterra competition to suggest that periodic perturbations might lead to a humped relationship between species diversity and either productivity or disturbance rate.

This analysis applies rigorously only to plant communities within a particular geographic region, communities which share a species pool and have similar levels of spatial heterogeneity. It thus does not consider the question of the evolution of diversity. It is tempting to suggest that at least a part of the latitudinal gradient in plant species diversity can be explained by this resource richness–species richness hypothesis. In comparison with tropical soils, the slow rate of decay of organic matter and the lower rainfall associated with temperate soils cause many temperate soils to be more nutrient-rich than many tropical soils (although there is considerable variation in both regions). This suggests that, given sufficient time, forests as diverse as some tropical rain forests (Ashton, 1977; Holdridge *et al.*, 1971; Fig. 42) may not occur in temperate regions because the soils of temperate forests are not sufficiently nutrient-poor. However, very diverse plant communities do occur in temperate regions, such as the Cayler prairie of Iowa.

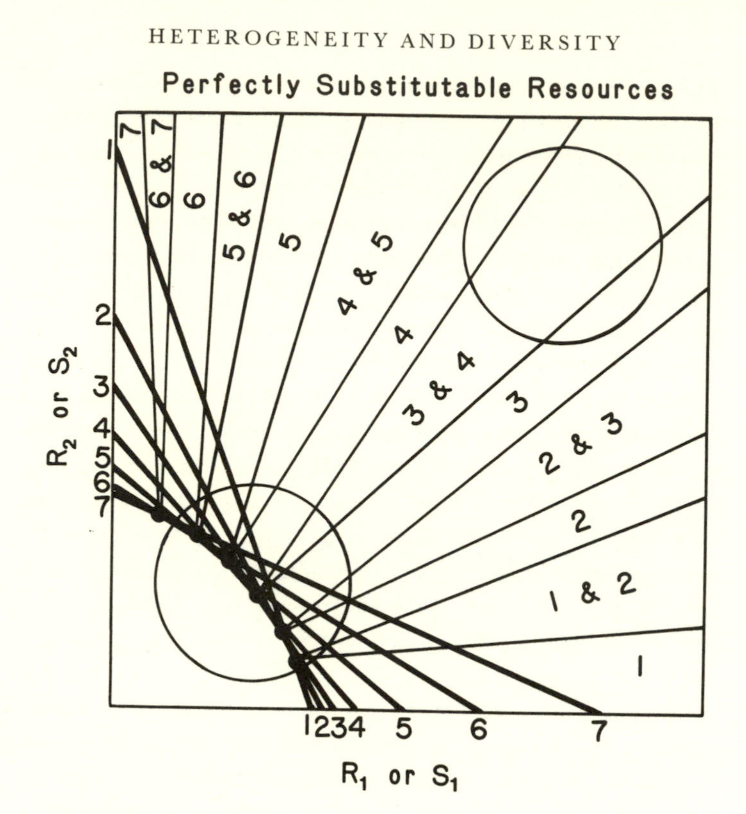

FIGURE 48. Seven species are shown competing for two perfectly substitutable resources. Because of the differences in their resource requirements, each of the species can stably exist in a habitat which has the appropriate proportions of the resources. As for essential resources, increased resource richness in a spatially heterogeneous habitat can lead to decreased species richness. However, the discussion of Chapter 9 suggests that the case illustrated above may be unlikely to occur.

This prairie, which occurs on very nutrient-poor sandy soil, has over 300 species of vascular plants (Werner and Platt, 1976).

Because the major limiting factors for plants (nutrients, light, water, open space) are inanimate (and thus not subject to natural selection), and because these are essential resources, plant-resource interactions are simple to model. A resource isocline approach may also be used for animal communities, but it would be more complex and less general. Animal resources fall into at least five classes, each represented by a differently shaped isocline (perfectly substitutable, complementary, antagonistic, switching, and hemi-essential). Although a plant community may have only two to four limiting resources, an animal community may have as many resources as there are consumers (Grubb, 1977). Even so, at equilibrium an animal community will have a series of equilibrium points, each with an associated region of coexistence. As for plant communities, enrichment could decrease the effective spatial heterogeneity, thus decreasing the number of species capable of coexistence in a spatially heterogeneous environment, as illustrated for perfectly substitutable resources in Figure 48. If, however, most animals have switching resources, as suggested earlier, a quite different result occurs, as will be discussed in Chapter 9.

SUMMARY

Contrary to the generalizations made by Grubb (1977), Connell (1978), and Huston (1979), this chapter has demonstrated that the diversity of plant communities need not be limited, in theory, by the common need of plants for a few nutrients. Differentiation in the optimal ratio of nutrients required by plants can account for the existence of rich, stable communities. Such differentiation could occur for several different pairs of nutrients. Theory predicts that plant community diversity should be maximal in moderately resource-poor habitats, and should decrease with either increases or decreases in resource richness. Many of the properties of natural and manipulated plant communities seem consistent with predictions made by this equilibrium theory of nutrient competition. The "humped" resource richness–species richness curves observed in several natural communities and the dramatic decreases in diversity following enrichment of both terrestrial and aquatic plant communities provide strong support for the model. The work of Grubb, Connell, Huston, and others suggests that the species diversity of plant communities may be understood in terms of the response of various species to disturbances. Chapter 8 considers this alternative hypothesis.

CHAPTER SIX

Resource Ratios and the Species Composition of Plant Communities

The previous chapters have suggested that the diversity of plant communities may be at least partially explained by a theory of resource competition in a spatially heterogeneous environment. This theory depends on the assumption that an individual plant species is a superior competitor for only a small range of resource supply ratios. If plants can only be superior competitors for a small range of resource supply ratios, spatial variation in resource supply ratios can lead to the stable coexistence of many more species than there are limiting resources. This chapter explores the possible validity of this assumption for both terrestrial and aquatic plant communities by looking at the dependence of the relative abundances of species on the ratio of various limiting resources. The scope of this chapter is limited to plant communities both because thorough data sets relating species composition to resource availability exist only for plant communities and because the theory being explored is expected, for reasons presented in Chapter 9, to apply universally to plant communities but to apply to only a limited number of animal communities.

In order to interpret many of the available data, it is helpful to consider some theoretical cases of the effects of resource ratios on plant community composition. Figure 49 shows four species competing for two limiting resources. The differences in the resource requirements and consumption characteristics of these species lead to their separation along the "resource ratio gradient" of Figure 49A. A resource ratio gradient is a convenient way to summarize how various rates of supply of resources will affect the outcome of competition. The line on Figure 49A which is labeled "Gradient" and which has eight points along it (labeled 1-8) is one such gradient. A resource ratio gradient is an arbitrary device which allows easy determination of the effects of various relative resource supply rates on competition. Points along this gradient range from supply of mainly R_1, to intermediate rates of supply of both resources, to supply of mainly R_2. Point 1 (see Fig. 49A) is a resource supply point in the region in which species A should competitively displace all other species, and point 8 is in the region in which species D should displace all other species. Figure 49B shows the approximate position along this gradient for each of these species in the *absence* of interspecific competition. In the absence of competition, each species reaches its peak population density at a different point along the resource ratio gradient. The maximum population density for each species occurs at the resource ratio at which the species is equally limited by the two resources, i.e., at $S_1/S_2 = c_{i1}/c_{i2}$. This may be called the optimum resource ratio for this species. The optimum ratio of each species is indicated with a broken line in Figure 49B and C.

In the *presence* of interspecific competition, the four species have a distribution along the resource ratio gradient as shown in Figure 49C. Each species reaches its peak population density in the presence of competitors at its optimum resource ratio. Figure 49 suggests that it may be possible to use the optimum resource ratios of various species as a way to predict how they should be distributed along a natural resource ratio gradient. Use of optimum resource ratios in place of the more complete representation of resource isoclines (as in Fig. 49A) is valid only if one critical assumption is met: the ZNGI's and consumption vectors of all the species must be related to each other as shown in Figure 49A. That is, each species must have at least one locally stable two-species equilibrium point. For such an

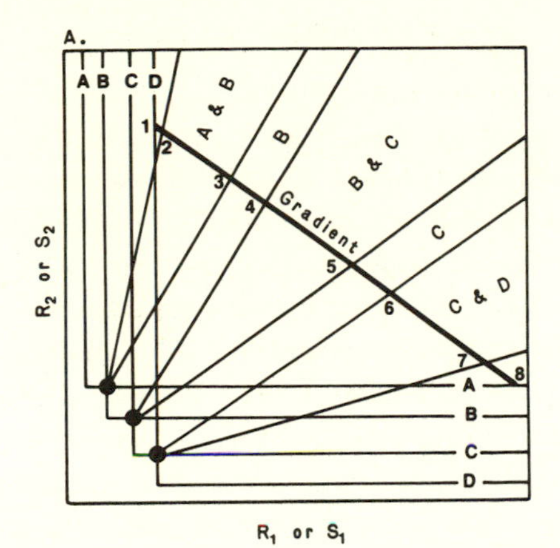

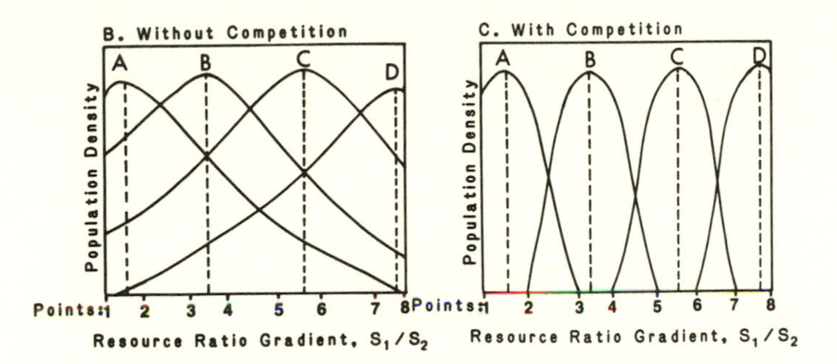

FIGURE 49. A. The ZNGI's of four species which differ in their resource requirements are shown along with the regions in which various species or pairs of species are predicted to be competitively dominant.

B. The qualitative pattern of the abundance of each of these species along the resource ratio gradient of part A of this figure *in the absence of competition* is shown. The distributions predicted by theory actually reach a pointed peak, not the rounded peak shown. A rounded peak is shown for simplicity and because spatial heterogeneity, which is expected in any natural community, would tend to round the peak. Although a mathematical purist may prefer peaked, not rounded curves, rounded curves will be used for the remainder of this book. (See, however, Fig. 77.)

C. The qualitative distribution of these four species in the presence of each other along the resource ratio gradient of part A. Note that each species reaches its peak population density at the ratio of resource availabilities for which it is equally limited by the two resources. This ratio, shown with a broken line in parts B and C of this figure, is the optimum resource ratio of each species.

equilibrium point to occur, the ZNGI of species with adjacent optimum resource ratios must cross, and the consumption vectors of these species must be such that each species consumes relatively more of the resource which limits its own growth rate at that two-species equilibrium point. This is comparable to saying that there must be tradeoffs in the abilities of species to compete for these two resources. The importance of this restriction is illustrated in Figure 50. Species *C* of this figure does not meet this restriction, for it does not have a locally stable two-species equilibrium point. Even though its optimum resource ratio falls between that of species *A* and species *B* (Figure 50B) it would not be able to coexist along a resource ratio gradient (Figure 50C). This cautionary statement is offered partly in response to some generalizations made by Rhee (1978) and Rhee and Gotham (1980). Thus, given that all species have at least one stable two-species equilibrium point, species which differ in their optimum resource ratio for the two limiting resources should separate along a resource ratio gradient. Separation is to be expected even when species compete in a heterogeneous environment, although the separation will be less sharp the greater the amount of spatial and temporal heterogeneity.

If differences in resource requirements are the mechanism allowing the coexistence of numerous species in spatially heterogeneous habitats, analysis of data for natural communities should show separation among species in relation to the ratios of limiting resources. Similarly, studies of the resource requirements of naturally occurring species should show that each species has a unique optimum resource ratio for the limiting resources. In addition, resource enrichment experiments performed on natural communities should show different species becoming dominant depending on the pattern of enrichment with limiting resources: the species which becomes dominant should be the species with an optimum resource ratio similar to the ratio at which resources are being supplied.

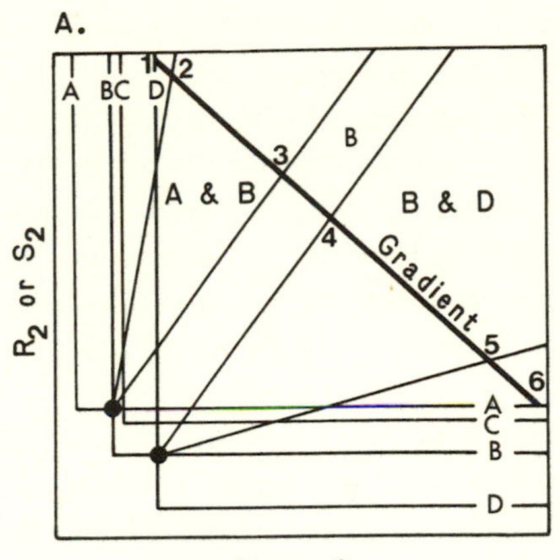

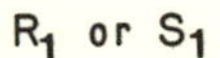

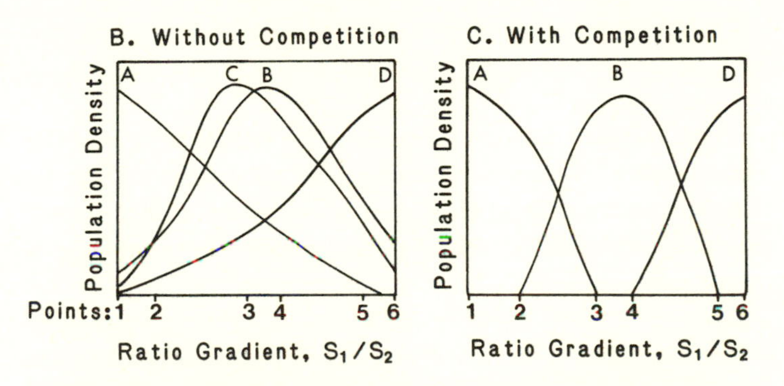

FIGURE 50. A. The ZNGI's of four species which differ in their optimum $S_1:S_2$ resource ratios are shown along with their predicted regions of dominance. Although species C has an optimum resource ratio which is intermediate between that of species A and B, it cannot coexist with these species along a resource ratio gradient because it has a higher requirement for both resources than the other species.

B. The qualitative pattern of the distribution of these four species along a resource ratio gradient in the absence of interspecific competition.

C. The qualitative pattern of the distribution of these four species along a resource ratio gradient in the presence of interspecific competitors. Note that species C is competitively displaced for all points on the gradient.

An ideal test of the ideas presented in this book would include all these sets of information on the same community, and thus allow multiple opportunities for the potential falsification of the theory of resource competition. Unfortunately, I know of no cases in which all of this information has been collected for a single community. However, various pieces of the picture have been obtained for a wide variety of plant communities. These examples will be reviewed in the rest of this chapter.

This review is not offered as "proof" of the validity of the theory developed in the preceding chapters. Only thorough experiments performed on a variety of natural communities will allow an ultimate judgment of its worth. Rather, the field and laboratory data seem sufficiently consistent with theory to suggest that further studies may be worthwhile.

RESOURCE REQUIREMENTS OF INDIVIDUAL SPECIES

Of all the information that has been collected on the resource requirements of individual species, the most relevant to this discussion are studies of the resource requirements of groups of plants occurring in the same habitat. Almost all of these studies have shown differences in the resource requirements of the species.

For instance, Rhee and Gotham (1980) have shown that eight species of freshwater algae differ in their optimal N:P ratio (Fig. 51). The optimal ratio ranges from 7 from the diatom *Melosira binderana* to 30 for the green alga *Scenedesmus obliquus*. Thus, if these species have ZNGI and consumption vectors which are related to each other as shown in Figure 49, *Melosira*, *Microcystis*, and *Synedra* should dominate low nitrogen but high phosphorus habitats, whereas *Fragilaria* and *Scenedesmus* should dominate high nitrogen but low phosphorus habitats. However, other work that has been done on the phosphate requirements of these species (Tilman, 1977; Tilman and Kilham, 1976; Rhee, 1978, 1974; Tilman, 1981; Holm and Armstrong,

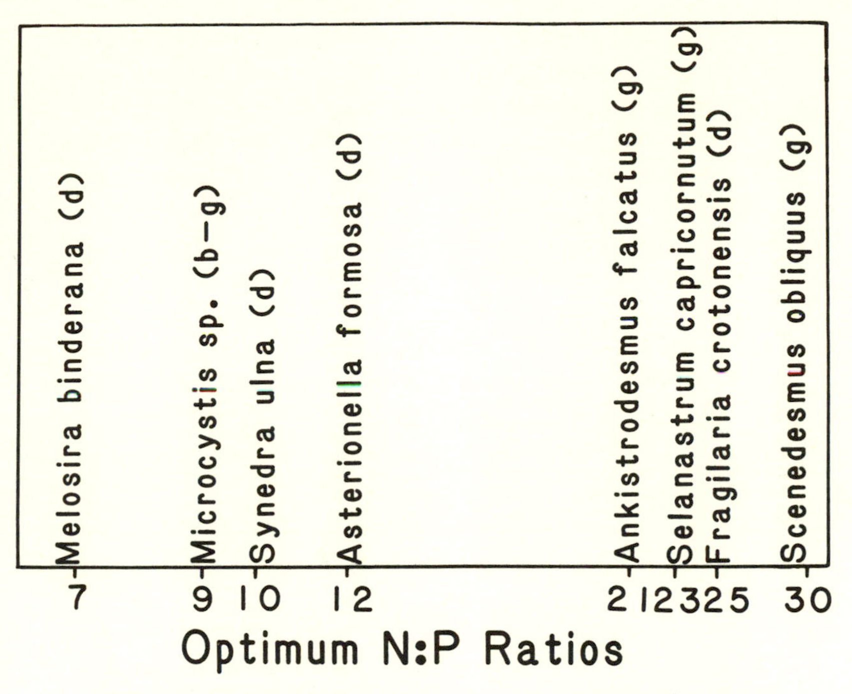

FIGURE 51. Optimum N:P ratios of several species of diatoms (*d*), green algae (*g*), and blue-green algae (*b-g*), as determined by Rhee and Gotham (1980). If these species have the necessary tradeoffs in competitive abilities for N and P, i.e., if the species are ranked such that a species which is a superior competitor for one resource is also an inferior competitor for the other resource compared to the adjacent species on the gradient, each of these species should dominate habitats with its optimum N:P ratio. N:P ratio is graphed on a log scale.

1981 suggests that some of the requisite two-species equilibrium points may not exist, at least at 20°C. For this reason, Rhee and Gotham (1980) must be interpreted cautiously.

Work by Tilman (1977, 1981) and Tilman and Kilham (1976) on the silicate and phosphate requirements of five species of freshwater diatoms allows them to be ranked in terms of their competitive ability along a Si:P ratio gradient (Fig. 52). The two entries for *Asterionella formosa* correspond to studies on two different clones of this species. Clone 1 was isolated from

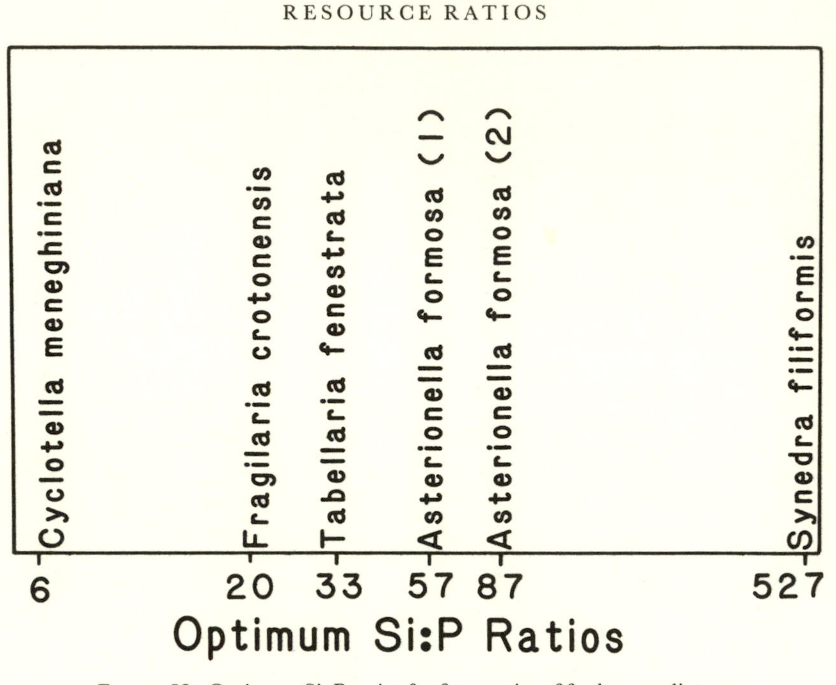

FIGURE 52. Optimum Si : P ratios for five species of freshwater diatoms (data from Tilman and Kilham, 1976; Tilman, 1981). At 20°C, only *Tabellaria fenestrata* lacks the necessary tradeoffs in resource competitive abilities for it to coexist with the other species. Si : P ratio is graphed on a large scale.

Lake Michigan, while clone 2 was isolated from small, eutrophic Frains Lake, Michigan. One of the five species, *Tabellaria fenestrata*, is an inferior silicate and phosphate competitor compared to the species adjacent to it along the Si : P ratio gradient at 20°C and is not able to coexist on this gradient (Fig. 33). The other species demonstrate the separation of four species of freshwater diatoms along an Si : P resource ratio gradient.

The resource requirements of six species of grasses have been reported in a series of papers by Bradshaw *et al.* (1960a, 1960b, 1958, 1964). For the minerals N, P and Ca, I used their data to estimate the relative competitive ability of each species for a given nutrient by using the percent of the maximal rate of

Limiting Resources

Competitive Ability	Nitrogen	Calcium	Phosphorus	Light
Best	Ns	At	Cc	Lp
↓	Cc	Ns	Ns	As
↓	At	Lp	Ac	At
↓	Ac	As	As	Cc
↓	Lp	Ac	At	Ac
Worst	As	Cc	Lp	Ns

Where: Ns = Nardus stricta
Cc = Cynosurus cristatus
At = Agrostis tenuis
Ac = Agrostis canina
Lp = Lolium perenne
As = Agrostis stolonifera

FIGURE 53. Estimated resource requirements of six grasses. The experimental studies of Bradshaw *et al.* (1958, 1960a, 1960b, 1964) were used to rank the grasses according to their estimated competitive abilities for the various limiting resources.

weight gain that occurred at the lowest level of the limiting nutrient. I assumed that the species which grew at the highest percent of its maximal rate was the superior competitor for that nutrient, that the next species was the second best competitor, etc. The data of Bradshaw *et al.* also allowed an estimation of the competitive ability of these species for limiting light. Because of the similarity of growth form of grasses, I assumed that increased weight would correspond with increased height, and that increased height would allow a species to be a superior competitor for light. Thus, the tallest (heaviest) species was assumed to be the superior competitor for light, etc. These estimated rankings of competitive ability for these four resources are shown in Figure 53. Examination of these rankings reveals that a species which appears to be the best competitor for one resource is often the worst competitor for another resource, and that competitive abilities of intermediate species switch with respect to each other. Such tradeoffs in resource requirements are required by theory if the mechanism allowing coexistence of these species is the specialization of each species on a different proportion of the limiting resources.

FIELD CORRELATIONS: NATURAL RESOURCE GRADIENTS

There have been numerous studies showing correlations between various plant resources and the species composition of plant communities. For aquatic communities, Pearsall (1930, 1932) was one of the first to suggest that different plankton communities occurred in response to different nutrient ratios, including N:P and (Ca + Mg)/(Na + K). The latter ratio has met continued criticism, but the former ratio has proven to be of some value (Schindler, 1977; Stoermer, Ladewski, and Schelske, 1978).

Figure 52 gave the optimum Si:P ratios for five algal species, of which four species (*Cyclotella*, *Fragilaria*, *Asterionella*, and *Synedra*) have been shown to have the tradeoffs needed for coexistence along a Si:P gradient. Laboratory competition experiments between various pairs of these species have generally supported the predicted competitive abilities of Figure 52 (Tilman 1977, 1981). We may ask, then, how the distribution of these species in nature depends on Si:P. The only available data of which I am aware are those of Kopczynska (1973), from her study of near-shore and river-plume algae in Lake Michigan. From this data set I have extracted the density of *Cyclotella meneghiniana*, *Fragilaria* sp., *Asterionella formosa*, and *Synedra* sp. in relation to the observed ratio of ambient SiO_2 to ambient PO_4. Figure 54A-D shows the relative abundance of each of these species versus Si:P. Note that, although there is considerable variance, *Cyclotella* reaches its greatest relative abundance at lower Si:P ratios, followed by *Fragilaria*, *Asterionella*, and *Synedra*. This pattern, summarized in Figure 54E, is in agreement with the optimal Si:P ratios of Figure 52. However, it is highly dependent on a few data points, and must be verified with additional field work.

These data may also be analyzed in a different manner. The optimal Si:P ratios of Figure 52 suggest that *Cyclotella* should be a superior competitor compared to any one of the other three species at low Si:P ratios. To determine whether Kopczynska's data were consistent with this prediction, I graphed the proportional abundance of each of the other species (of the total of that species and *Cyclotella*) against the Si:P ratio. Figure 54F shows that *Fragilaria* is very rare compared to *Cyclotella* at low Si:P ratios, and is increasingly relatively abundant at higher Si:P ratios. The observed trend is highly statistically significant. Similarly, *Asterionella* is rare compared to *Cyclotella* at low Si:P and is more abundant at high Si:P. *Synedra* follows the same pattern. In all three cases, the qualitative pattern of species dominances along a natural Si:P gradient is consistent with theory, and all three patterns are highly significant statistically.

The actual Si:P ratio at which the transition occurs from dominance by *Cyclotella* to dominance by another species does not seem consistent with the data of Figure 52. Figure 54F-H shows that this transition begins at ratios of ambient silicate to ambient phosphate of about 60, not the predicted ratio of 6. However, the data of Figure 52 predict a transition from *Cyclotella* to one of the other three species at a Si:P *supply* ratio of 6, but make no prediction concerning ratios of *ambient* concentrations. Use of ambient ratios assumes that there is a relationship between ambient Si:P ratios and ratios of Si:P supply rates. Because phosphate is regenerated much more rapidly in the water column than is silicate, ambient Si:P ratios are likely to greatly overestimate supply ratios. If

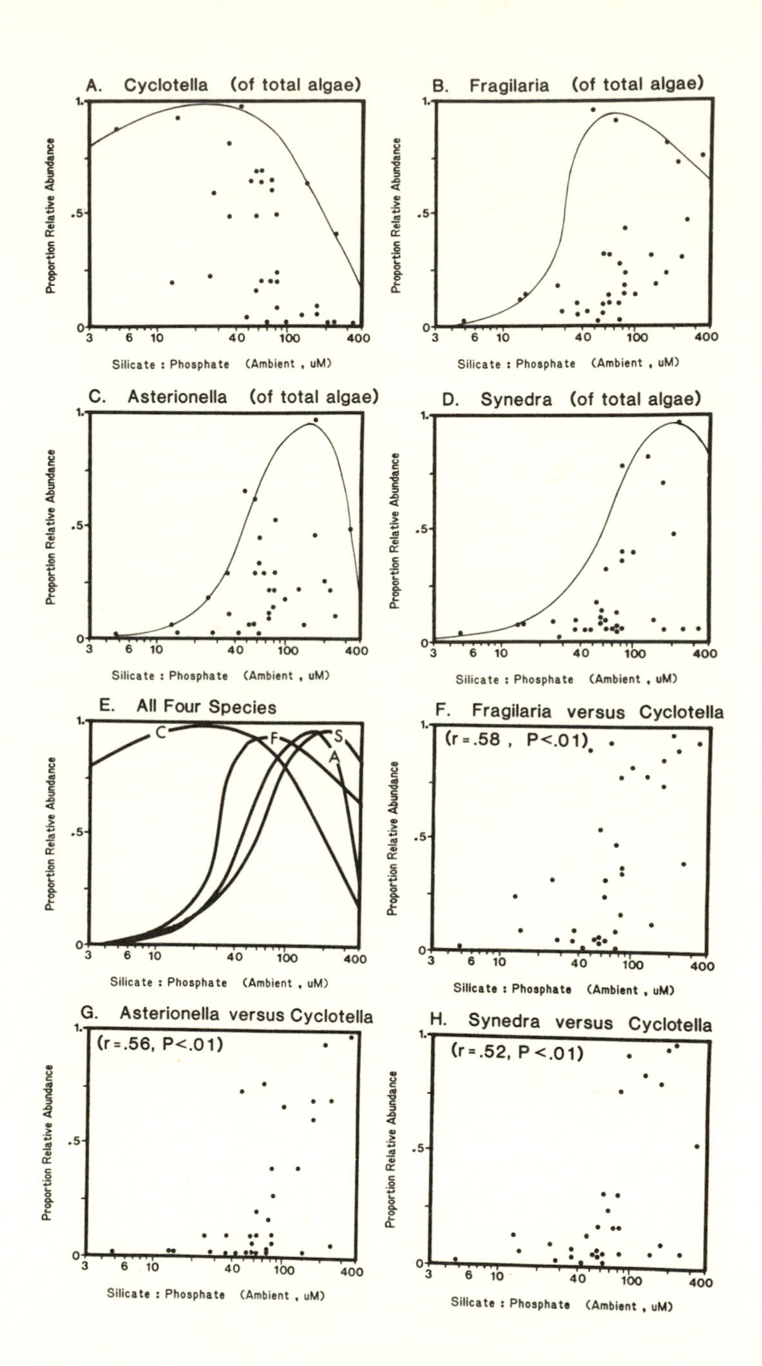

A. Cyclotella (of total algae)
B. Fragilaria (of total algae)
C. Asterionella (of total algae)
D. Synedra (of total algae)
E. All Four Species
F. Fragilaria versus Cyclotella
(r = .58 , P<.01)
G. Asterionella versus Cyclotella
(r = .56, P<.01)
H. Synedra versus Cyclotella
(r = .52, P <.01)
C
F
S
A
Proportion Relative Abundance
Silicate : Phosphate (Ambient , uM)
1.
.5
0
3
6
10
40
100
400

FIGURE 54. The relationship between ambient Si : P ratios and the relative abundance of four Lake Michigan diatoms. Relative abundance is expressed, for parts A–E, as the absolute abundance of a species divided by the total algal abundance, and scaled so that the peak abundance of a species is 1.0. Parts F, G, and H show the relationship between ambient Si : P ratios and the relative dominance of *Fragilaria*, *Asterionella*, or *Synedra* compared to *Cyclotella*. This was calculated as the density of one of these species divided by the sum of the density of that species and the density of *Cyclotella*, with density expressed as cells/ml. See also Figures 33 and 34. The data in part G are a correction of a multiplication error which occurred in preparing Figure 14 of Tilman (1977). In converting ppm SiO_2 and ppb PO_4-P in Kopczynska's (1973) data set into μM Si and μM P, a consistent 3 × multiplication error was made. This shifted all of Kopczynska's data toward apparently lower ambient Si : P ratios. Part G above thus should be substituted for Figure 14 in Tilman (1977), which was incorrect. I apologize for this error, and thank Dan Sell, Heath Carney, and Gary Fahnenstiel of the University of Michigan for bringing it to my attention.

phosphate were regenerated 10 times more rapidly than silicate, an ambient Si:P ratio of 60 would be comparable to a supply ratio of 6. Until much more work is done on the processes controlling internal nutrient cycling, we will have to use ambient resource ratios as an index of resource supply ratios, and we will be unable to determine the quantitative goodness of fit of field data to theory. However, the data in Figure 54F-H are qualitatively consistent with predictions based on the data of Figure 52.

For terrestrial plant communities, there have been a variety of studies which have shown separation of species in relation to various factors, including elevation (Whittaker and Niering, 1975; Hanawalt and Whittaker, 1976, 1977a, 1977b), moisture (Beals and Cope, 1964; Zedler and Zedler, 1969), and nutrients (e.g., Ashton, 1977; Beadle, 1966, 1954; Box, 1961; Coaldrake and Haydock, 1958; Kruckeberg, 1954; McColl, 1969; Moore, 1959; Pigott and Taylor, 1964; Shields, 1957; Snaydon, 1962; Van Den Bergh, 1969; Wagle and Vlamis, 1961). As noted by Hanawalt and Whittaker (1976, 1977a, 1977b), and by many others, many of the factors on which species separation is observed are themselves correlated with other factors. Thus, although such correlational studies may seem to suggest that a particular pair of factors may be important for the ecology of a group of species, only experimental studies on natural communities will be able to separate spurious correlation from causation.

N:P RATIOS AND ALGAL DOMINANCE PATTERNS

Kilham and Kilham (1981) have reviewed several published cases of the relationship between nutrients and the species composition of freshwater phytoplankton communities. One of their more interesting cases is that of Lake 227 in the Experimental Lakes Area (E.L.A.) of Canada. Schindler (1977) reported that this lake was dominated by the green alga

Scenedesmus when the ratio of nitrogen to phosphorus being added to the lake was 31 : 1. When the pattern of fertilization was changed to give a loading ratio for N:P of 11 : 1, the lake became dominated by *Aphanizomenon*, a blue-green alga. Interestingly, Rhee (1978) reported that the optimum N:P ratio of *Scenedesmus* is 30 : 1, amazingly close to the ratio at which this species is dominant in an experimentally manipulated lake. Lake 227 may be compared with another nearby lake, 226NE, which is similar in its physical characteristics to lake 227 except that lake 226NE was manipulated to give a nutrient loading ratio for N : P of 12 : 1. Lake 226NE was dominated by the blue-green alga *Anabaena*. Both of these lake manipulation studies revealed that blue-green algae became dominant at low N:P.

Low ratios of nitrogen to phosphate are conditions in which the nitrogen fixation capabilities of blue-green algae are likely to give them a competitive advantage. Figure 55 shows hypothetical ZNGI and consumption vectors for a nitrogen fixing organism, such as a blue-green alga, and for a non-fixer. It assumes that the nitrogen fixer, because of its ability to reduce N_2 to NH_3^+, will have a lower N requirement than the non-fixer, but that the non-fixers will have a lower phosphate requirement than the fixers. This would lead to these two groups having a distribution along a N:P gradient as illustrated in Figure 55B.

Val H. Smith (unpublished manuscript) has obtained from a literature review the relationship between N:P ratios in lakes and the percent of the total phytoplankton that are blue-green algae (Fig. 56). Curves for Lake Washington (Seattle, Washington) from studies by W. T. Edmondson show that blue-green algae dominated this lake during the time that total nitrogen (TN) to total phosphorus (TP) ratios were below about 24. However, with sewage diversion controlling phophorus inputs to the lake, the TN:TP ratio increased, and an increasingly great portion of the algae in the lake were the non-nitrogen fixing diatoms and green algae (Fig. 56A). Smith noted that

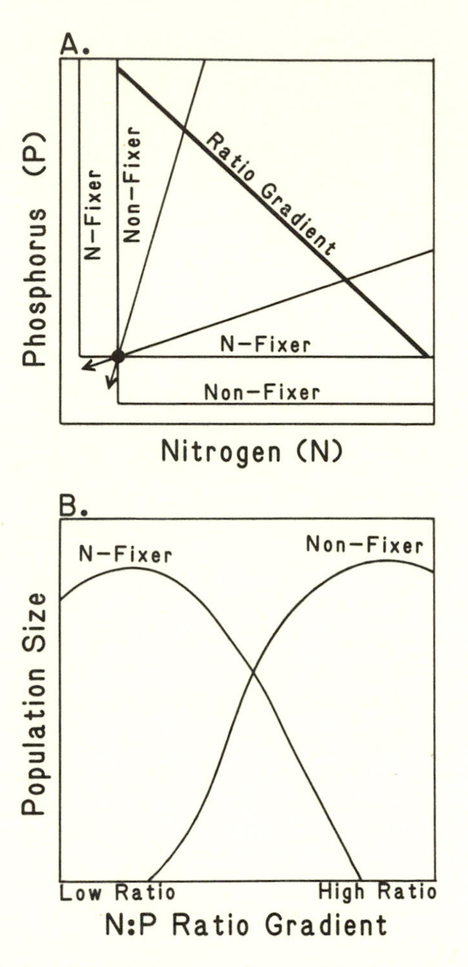

FIGURE 55. A. A hypothetical case of competition between a nitrogen-fixing and a non-nitrogen-fixing species. This assumes that the nitrogen fixer is an inferior competitor for P compared to the non-fixing organisms. Although nitrogen fixers need not be inferior competitors for phosphorus, the work of Zevenboom *et al.* (1981; see Fig. 18) suggests that nitrogen fixation does increase the requirements of a species for another resource.

B. The qualitative pattern of abundance of these two species along an N:P resource ratio gradient.

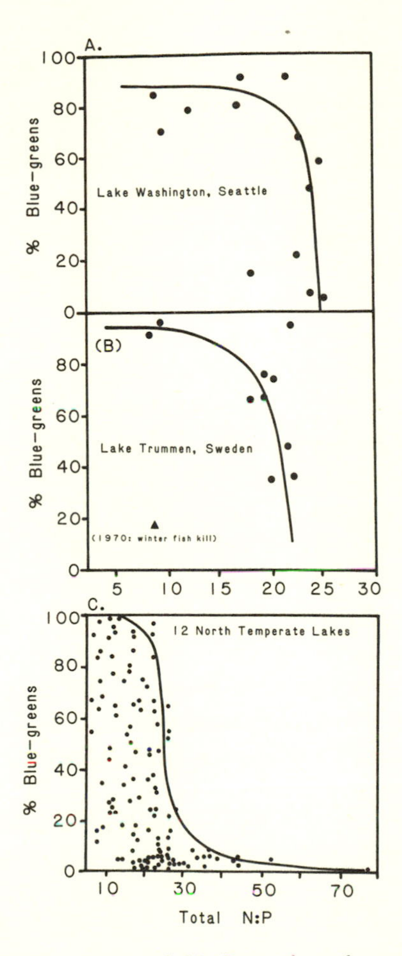

FIGURE 56. The three parts of this figure show the percent relative abundance of blue-green algae in several different lakes in relation to the N:P ratio of the lake. Note that blue-green algae, many species of which are capable of nitrogen fixation, tend to be increasingly dominant in lakes with low N:P ratios. This is so for Lake Washington, Seattle (part A), Lake Trummen, Sweden (part B), and for a summary of these two lakes and ten other north temperate lakes. Data were kindly provided by Val H. Smith.

the lake was dominated by the blue-green alga *Oscillatoria* spp. before sewage diversion, at which time the lake N:P ratios approximated the optimal N:P ratio for this species. The data for Lake Trummen, Sweden, (Fig. 56B) show a similar relationship between the TN:TP ratio and the percent dominance by blue-green algae. There is one intriguing data point—the datum for 1970—which deviates from the curve shown. The winter before the 1970 growth season there was sufficiently thick ice and snow on Lake Trummen that the water became anoxic, causing the death of most of the fish in the lake. Because of this fish kill, predation rates on herbivorous zooplankton were very low, allowing high population densities of zooplankton to develop. Contrary to the generalizations made by Porter (1973, 1976), high rates of grazing by herbivorous zooplankton led to unusually low population densities of blue-green algae in 1970. As an alternative to the hypothesis of differential susceptibility to zooplankton predation offered by Porter for her work on blue-green and green algae, I might suggest that the major effect of high zooplankton levels is on the rates of supply of nutrients. If it is found that zooplankton grazing tends to regenerate nitrogen more quickly than phosphorus, the results of the shift in the abundance of algal groups in Lake Trummen would be consistent with the hypothesis of differential nutrient regeneration by herbivores. Smith noted that the dominant alga in Lake Trummen when it had low N:P ratios was *Microcystis* spp., which has an optimal N:P ratio of 4.1, very close to the ratio of available N:P in the lake at that time.

In addition to the data from Lakes Washington and Trummen, Smith similarly compiled data from twelve lakes worldwide. Although there is considerable scatter when data from so many lakes are combined, Figure 56C shows a dramatic effect of N:P ratios on the taxonomic composition of these lakes. Blue-green algae tend to dominate only those lakes that have a TN:TP ratio less than about 25. Lakes with TN:TP greater than about 25 are dominated by the non-nitrogen fixing diatoms and green algae. Smith believes that some of the scatter in Figure 56C can be explained by competition for light.

N:P RATIOS AND TERRESTRIAL PLANTS

Many legumes (peas, acacias, etc.) are capable of fixing nitrogen because of a mutualistic interaction with bacteria. We thus might expect legumes to be most dominant in sites which are low in nitrogen but relatively high in other resources (light and nutrients). Such a pattern has been repeatedly observed in correlational studies. For instance, Campbell reported in 1927 that "wild legumes compete with other species best on poor soils, the percent legumes decreasing progressively as soil nitrogen increases." Similar trends were reported by Young (1934) for natural vegetation of the Adirondack Mountains. Foote and Jackobs (1966) reported that partridge pea (*Cassia fasiculata*), a nitrogen fixer, is more common on soils low in nutrients, especially nitrogen, but only in areas that had been recently disturbed. Such disturbed sites provide a high light environment. These observations suggest that *Cassia fasiculata* may be a good competitor for nitrogen and a poor competitor for light.

The fertilization experiments performed at Rothamsted, England, for the last 120+ years provide experimental data on the effect of nitrogen on the dominance of legumes. Of the 20 experimental plots, 14 have received the same fertilization treatment for the entire period. These may be divided into two groups: those receiving no nitrogen and those receiving nitrogen. Five plots (Plots 3, 4^1, 7, 8, and 12) were either never fertilized or fertilized with elements other than nitrogen. Nine plots (Plots 1, 4^2, 9, 10, 11^1, 14, 16, and 17) were fertilized with nitrogen and often other mineral elements. Figure 57

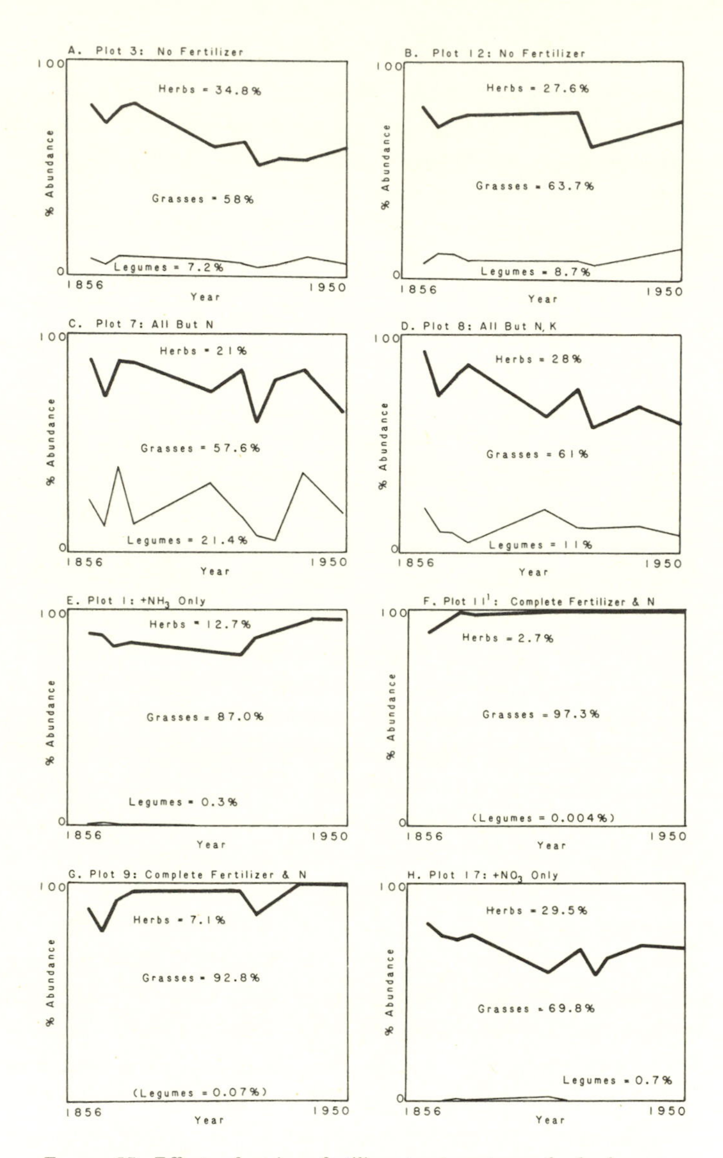

FIGURE 57. Effects of various fertilizer treatments on the herb, grass, and legume composition of some of the Rothamsted plots. The distance from the top of the graph to the thick line is the percent abundance of herbs in the plots, expressed on a weight basis; the distance from the thick line to the thin line is the percent composition of grasses; and the distance from the bottom of the graph to the thin line is the percent composition of legumes in the plot. The average percent composition of each plant group over all samples is given in each figure, and the fertilizer treatment is noted at the top of each figure.

shows the percent composition of legumes, grasses, and miscellaneous herbs in some of these plots. Note the dramatic difference in percent legumes in the plots receiving no nitrogen compared to those receiving various forms and amounts of nitrogen. Using the non-parametric Wilcoxon's two-sample test (Steel and Torrie, 1960), the plots receiving no nitrogen compared with those receiving some form of nitrogen were found to differ significantly ($P < 0.01$) in the percent legumes, percent grasses, and percent miscellaneous herbs, with the plots receiving nitrogen having a greater percent grass composition and a lower percent composition of legumes and herbs. The Wilcoxon test also revealed a significant decrease in the number of species of legumes, grasses, and miscellaneous herbs in the plots receiving N compared to those receiving no N ($P < 0.01$).

The two unfertilized plots (3 and 12) may be compared with Plot 7, which received all mineral elements except nitrogen (compare Fig. 57A and B with C). Because of its treatment, most plants in Plot 7 should be limited by nitrogen. The percent legumes in this plot should be higher than in the controls. Wilcoxon's test revealed a significantly higher ($P < 0.01$) percent legumes and a significantly lower ($P < 0.01$) percent miscellaneous herbs in Plot 7 compared to the unfertilized plots, with no significant difference in the percent composition of grasses. The number of species of legumes increased, but not significantly, in Plot 7, and there was a significant ($P < 0.01$) decrease in the number of miscellaneous herb species in Plot 7 compared to the controls.

Another comparison comes from Plots 14 and 16. Both received complete mineral fertilizers, but Plot 16 received half the amount of nitrogen supplied to Plot 14. Plot 14 averaged 1.4% legumes, while Plot 16 averaged 7.1% legumes. These are significant differences ($P < 0.05$, Wilcoxon's test). The number of species of legumes and the number of species of miscellaneous herbs in the high nitrogen plot were significantly lower than the number in the low nitrogen plot ($P < 0.05$,

Wilcoxon's test), with the number of species of grasses being unaffected by the treatment.

Inspection of Figure 57 reveals that the major effect of nitrogen fertilization on legume composition seems to have occurred between 1856, when the plots were first fertilized, and 1862, when the plots were first sampled for botanical composition. Although it seems unlikely, the possibility exists that the plots differed initially in their percent composition of legumes, and that the data just discussed might thus represent initial differences between the plots, and not the effect of nitrogen fertilization. This possibility is refuted by other experiments done at Rothamsted. Figure 58 shows changes in the percent composition of plots which were initially fertilized with nitrogen, and then observed after nitrogen fertilization was halted. Plot 5 was fertilized with ammonia from 1856 until 1897. During that period, its legume composition averaged 0.3%. In 1897, it was split into two subplots. Plot 5^1 has been unfertilized from 1897 on, while Plot 5^2 has received all nutrients *except* nitrogen since 1897. As shown in Figure 58, the percent legumes in these plots increased, with a much more rapid increase in Plot 5^2. This is consistent with the theory presented in Figure 55, because Plot 5^2 received all nutrients except nitrogen, thus driving all the species in the plot in the direction of nitrogen limitation. A similar experiment was performed on Plot 6, which was fertilized with ammonia from 1856 until 1868, during which period its legume composition averaged 0.2%. After 1868, it was fertilized with all nutrients except nitrogen, and its legume composition averaged 19.4%. Plot 15 was fertilized with sodium nitrate from 1856 until 1875, after which time it was fertilized with all mineral elements except nitrogen. It averaged 0.07% legumes during the time of nitrate fertilization, and 20.0% legumes during the time of fertilization with all elements except nitrogen.

The results summarized above show strong effects of resource availability on the relative dominance of major taxonomic

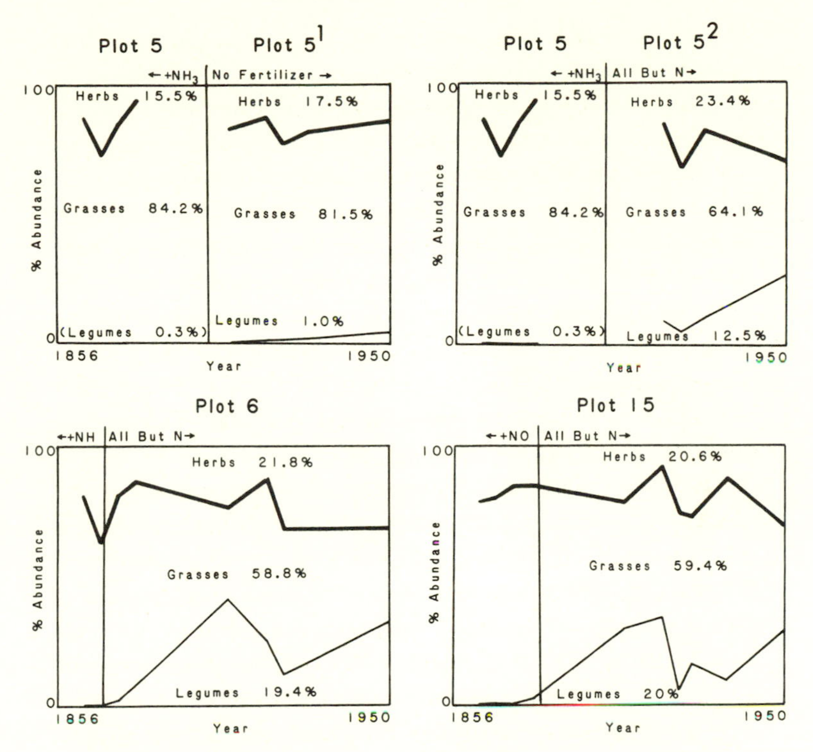

FIGURE 58. These four plots received one fertilizer treatment for a period of time, and then a different treatment. When the plots received nitrogen as either ammonia or nitrate, legume composition decreased to very low levels. Legume composition increased when fertilization was stopped (Plot 5^1), especially if all nutrients except N were applied (Plots 5^2, 6, and 15). All data are from the Rothamsted plots, and the notation used is the same as for Figure 57.

groups in the Rothamsted Park Grass Experiments, and provide the most complete experimental tests of such effects performed to date. Although the trends for taxonomic groups in Figures 57 and 58 are consistent with qualitative predictions of a resource-based theory of plant competition, the theory is designed to make predictions at the species level. How, then is the species composition of the various Rothamsted plots influenced by the pattern of fertilization? The following pages explore this question in increasingly great detail, starting first with a discussion of the long-term effects of fertilization on species composition. This is followed by an examination of the dynamics of competition in the plots, and then by an analysis which attempts to explain the results of the Rothamsted experiments in the context of the resource-based approach developed in this book.

Because of the strong effects of phosphorus and especially nitrogen on the productivity and species composition of the Rothamsted plots (Brenchley and Warington, 1958), let us first explore the long-term effects of different levels of these nutrients on the species composition of the plots. This can be done by combining information on the soil chemistry of the plots (Warren and Johnston, 1963) with information on the species composition attained in the plots after almost 100 years of fertilization (Brenchley and Warington, 1958). Although Warren and Johnston did not measure *in situ* rates of mineralization of nitrogen and phosphorus, they did measure the total amount of nitrogen and the total amount of phosphorus in the soils. The ratio of total nitrogen (TN) to total phosphorus (TP) may be used as an index of the relative rates of supply of these two frequently limiting nutrients. I assume that the TN and TP of 1959 are representative of the preceding years, and thus look for relationships between TN:TP and the species composition of the plots during the last census period for which data are available, 1948–1949. Because of the strong effect of pH < 4.0 on diversity patterns in these plots (see Fig. 45), I will only consider the 13 plots which had soil pH in the range from 4.2 to 6.0 (Plots 3U, 12U, 4^1U, 8U, 7U, 4^2L, 10L, 9L, 11^1L, 11^2L, 14U, 16U, and 17U, where U = unlimed subplots and L = limed subplots). In these plots, there were 9 species that were 8% or more of the total biomass in at least one of these plots. Figure 59 shows the relative abundances of each of these 9 species graphed against soil TN:TP. For 7 of the species, there seems to be a rather narrow range of TN:TP

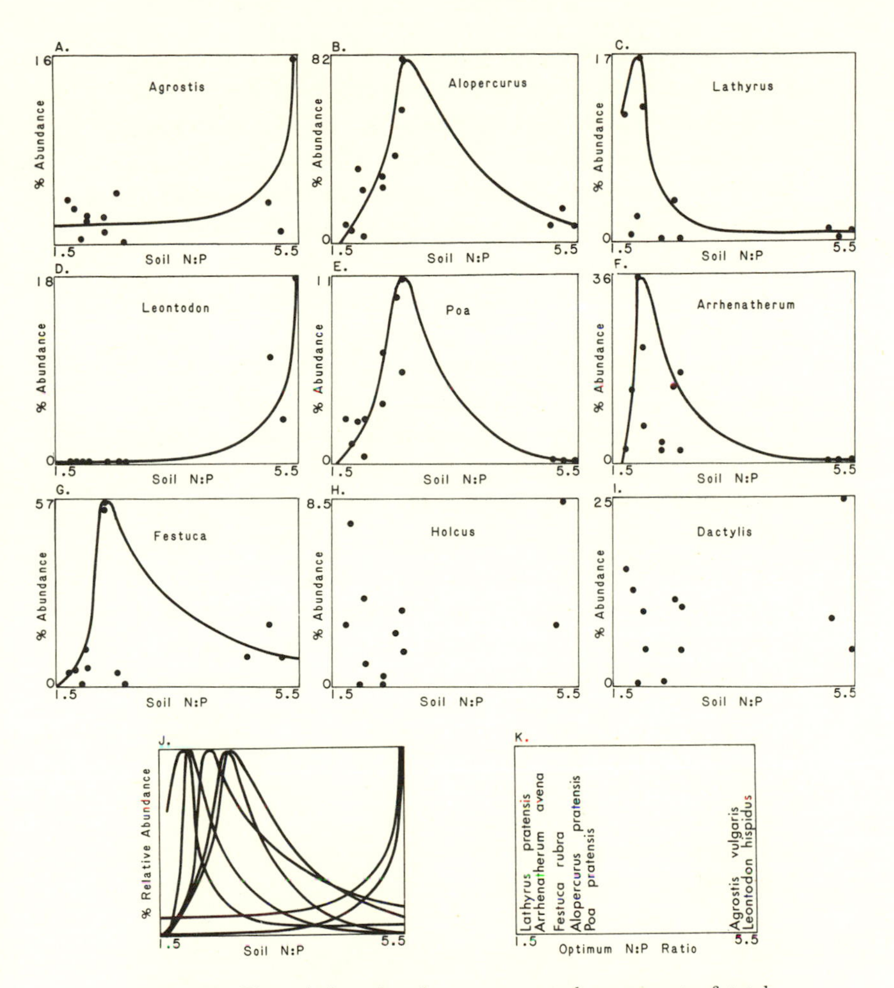

FIGURE 59. The relative abundances, expressed as percent of total community by biomass, of nine common plants in the Rothamsted plots are graphed against soil N:P ratios. The ratio used is a ratio of total N to total P, as given in Warren and Johnston (1963). Because of the effect of soil pH on plant community composition, I have included only those plots with a pH from 4.2 to 6.0. Parts J and K summarize the curves given separately in parts A to I.

ratios at which the species is dominant. For two grass species, *Dactylis* and *Holcus*, there is no apparent dependence of relative abundance on the TN:TP ratio. The curves, hand-drawn to each data set, are shown in summary in Figure 59J. There is an apparent separation of these 7 species along the TN:TP ratio gradient. Note that a legume (*Lathyrus pratensis*) is dominant at very low TN:TP ratios, that a herb (*Leontodon hispidus*) is dominant at high TN:TP ratios, and that grasses are dominant at intermediate ratios.

Figure 59 illustrates the relationships between species dominance and N:P ratios which existed after almost 100 years of fertilization in the Rothamsted plots. The population dynamics of some of these species in some of these plots are illustrated in Figures 60, 61, and 62. Figure 60 shows the changes in the species composition of an unfertilized plot (Plot 12) during the 90-year period for which data are available. Although there were changes in the relative and absolute abundances of the 8 dominant species shown (another 30 species are omitted for graphical clarity), the species composition of the plot remained relatively constant. Throughout the period, *Festuca* remained dominant and *Arrhenatherum* remained the least abundant of the 8 species. The other species, though they fluctuated, tended to remain at intermediate abundances.

The relative constancy illustrated in Figure 60 for an unfertilized plot may be compared with the dramatic changes in the abundances of these same 8 species in Plots 14 and 16, which received a complete mineral fertilizer with nitrogen supplied as nitrate (NO_3). Plot 16 (Fig. 61A) received nitrate at half the rate of Plot 14 (Fig. 61B). The dynamics of competition in Plot 16 show a strong tendency for two species which were subdominants in the unfertilized plots to become dominant with fertilization. *Arrhenatherum* increased in biomass by almost 3 orders of magnitude following fertilization, to become a co-dominant with *Alopecurus*, which itself increased in biomass by a factor of about 50. *Agrostis* and *Festuca*, which dominated the

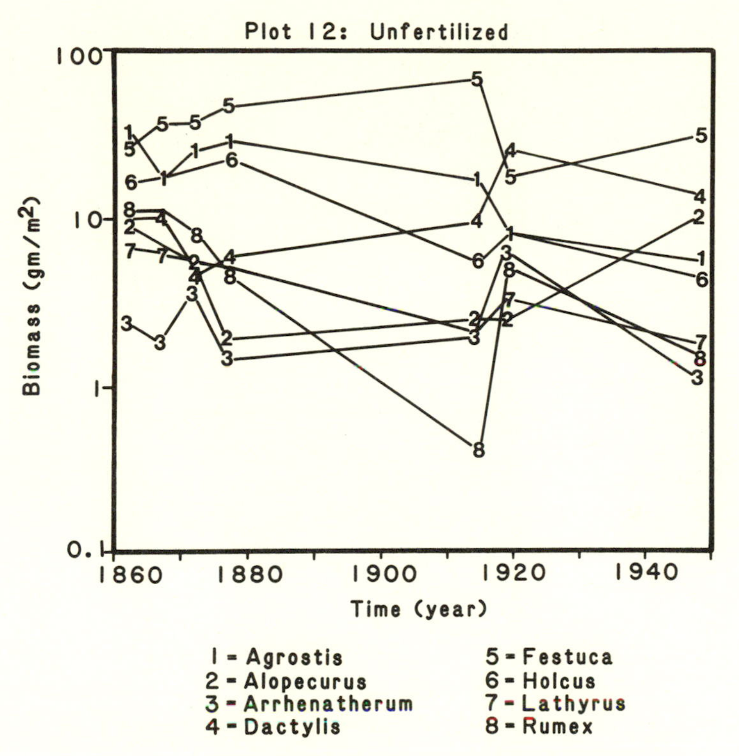

FIGURE 60. The dynamics of community change in an unmanipulated Rothamsted plot (Plot 12). The biomass of eight dominant species throughout all sampling dates is given. Note the constancy of this community compared to the plant communities which were subject to fertilization (Figs. 61 and 62).

plot at the start of fertilization, both decreased dramatically in abundance. *Arrhenatherum* and *Alopecurus* became even more dominant in Plot 14, which received more nitrate. Four of the remaining six species were driven to virtual extinction. Thus, complete fertilization with nitrate led to a low-diversity community dominated by two species.

A very similar pattern occurred in the plots receiving complete mineral fertilization but with ammonia as the nitrogen source (Fig. 62). These plots were dominated by a single species,

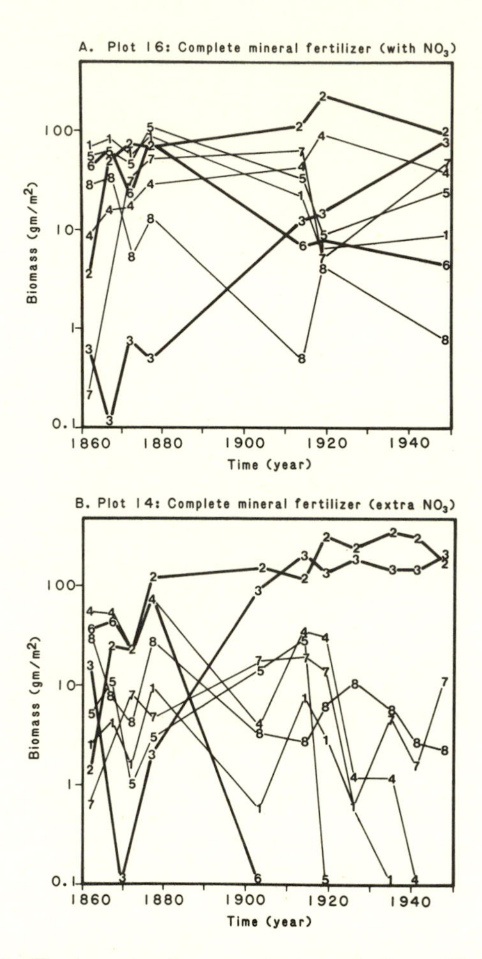

FIGURE 61. The dynamics of community change in the two Rothamsted plots receiving complete mineral fertilization with nitrate as the nitrogen source are shown for the same eight species illustrated in Figure 60. The species are numbered as indicated in Figure 60. *Alopecurus* (2) and *Arrhenatherum* (3) tend to dominate Plot 16 and are much more dominant in Plot 14, which received added nitrate. The complete mineral fertilization of Plot 14 led to a two-species community.

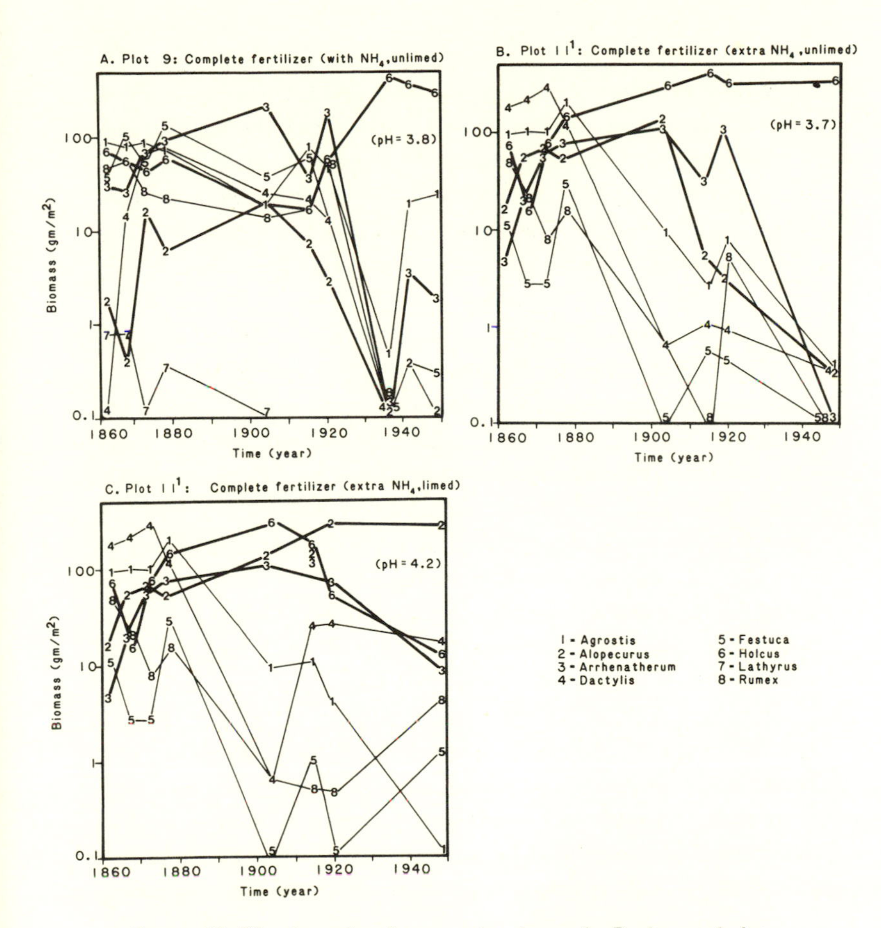

FIGURE 62. The dynamics of community change for Rothamsted plots receiving complete mineral fertilizer with nitrogen supplied as ammonia. There is a strong tendency for *Holcus lanatus* (6) to dominate the unlimed plots, especially Plot 11^1 which received more ammonia than Plot 9. The addition of lime to a subsection of Plot 11^1 raised the pH to 4.2, leading to a replacement of the dominant species, *Holcus lanatus*, by *Alopecurus pratensis* (2). *Alopecurus pratensis* also dominated Plot 14 (Fig. 61), which received nitrate and had a pH of 6.0. See also Figures 63 and 66.

with many other species being driven toward extinction. The two plots which were not limed and which thus experienced the soil pH decrease associated with ammonia fertilization were dominated by *Holcus lanatus*. The plot receiving lime was dominated by *Alopecurus*. Although there are considerable fluctuations, the data of Figure 62A and B suggest that *Holcus* is more dominant in the plot receiving ammonia at the higher rate.

Figures 61 and 62 make several important points. First, the rate of competitive displacement in these manipulated communities is relatively slow. Although the trends toward increasing dominance by *Alopecurus* and *Arrhenatherum* in Plots 14 and 16 were clear from an early stage in the experiments, the full effect of the imposed pattern of fertilization was apparently not realized until almost 100 years after fertilization for Plot 14. The data for Plot 16 suggest that that community may not have reached a new equilibrium after almost 100 years of fertilization. Such apparently slow rates of competitive displacement may seem surprising to those who supposed that, by virtue of their size, the species dominant in the Rothamsted plots were annuals or short-lived perennials. On the contrary, most of these species are long-lived perennials which often reproduce clonally via underground tissues. Clones of the various species may still be surviving from the period before the Rothamsted experiments were started. Some clones of plant species may age in the thousands of years (Harper, 1977). The rate of competitive displacement, when considered in terms of the life span of the organisms involved, may be seen to be at least as rapid as that observed for such short-lived organisms as algae, for which competitive displacement may require 10 to 20 generations (30 or more days).

A second point made by the data of Figures 61 and 62 is that complete mineral fertilizer leads to dominance by one or two species, with all other species being driven to very low population densities and even to extinction. This effect is increasingly pronounced in the plots receiving the most nitrogen. These are the plots which attained the highest productivity and total biomass. In the plots which received complete mineral fertilization with added nitrogen, it seems likely that almost all of the species were limited by the same resource, light. It is interesting to note that these communities became dominated by one or two species and that most other species were driven to extinction.

A third point made by these figures is the dependence of the identity of the dominant species following complete mineral fertilization on the soil pH. Complete fertilization with nitrate, which causes a moderate increase in soil pH, leads to dominance by *Alopecurus* and *Arrhenatherum*, whereas complete fertilization with ammonia, which causes acidic soils (pH less than 4.0), leads to dominance by *Holcus*. This effect of soil pH on dominance patterns is reinforced by the results of experiments in which a subsection of a plot receiving ammonia was limed. For instance, as shown in Figure 62C, liming, which raised pH from 3.7 to 4.2, led to dominance by *Alopecurus* in a plot which was completely dominated by *Holcus* in the unlimed section (Fig. 62B).

The dependence of the relative abundance of *Holcus*, *Alopecurus*, and *Arrhenatherum* on soil pH for plots receiving mineral fertilization with ammonia as the nitrogen source (including both limed and unlimed subsections of plots) is shown in Figure 63. *Holcus* reaches its maximum abundance in plots with soil pH less than 4.0, *Alopecurus* reaches its greatest dominance in plots with soil pH of about 4.3, and *Arrhenatherum* reaches its peak abundance for plots with a soil pH of about 4.7. Thus, these three species, which are the three species which dominate the plots receiving complete mineral fertilization and are thus probably the superior competitors for light, differ in their optimal pH. This difference in optimal pH is analogous to the difference in optimal temperature discussed earlier for two algal species competing for limiting silicate (Fig. 16). Although the species do not compete for pH, pH influences

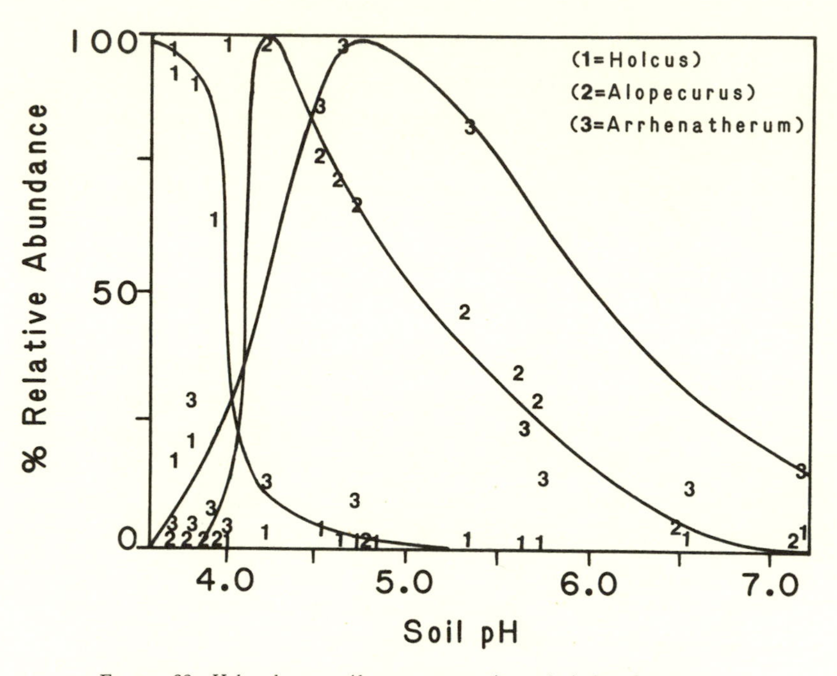

Figure 63. *Holcus lanatus*, *Alopecurus pratensis*, and *Arrhenatherum avenaceum* are the three species which tend to dominate the Rothamsted plots which have received complete mineral fertilization. Each of these species, however, tends to dominate only those plots which fall within a certain pH range, as the data from the Rothamsted experiments shown above demonstrates. These data are from those plots which received nitrogen as ammonia, and include both the limed and unlimed sections of these plots for the two sampling dates for which Warren and Johnston (1963) provided pH values. The percent relative composition for each species comes from the nearest available sampling date to the time of pH determination. The distributions for each species have been standardized to represent the highest abundance reached by each species as 100%. The observed maximal dominances of these species were: *Holcus*, 99.9%; *Alopecurus*, 82%; and *Arrhenatherum*, 17.5%.

their reproductive rate, and is thus an important physical factor determining the distribution and abundance of the species.

Although the analysis is complicated by changes in soil pH, as the preceding discussion illustrates, the data from the various fertilization experiments performed at Rothamsted can be analyzed in terms of the isocline approach to resource competition developed in this book. To control partially for pH effects, let us consider, again, those plots receiving nitrogen as ammonia, and let us exclude those subplots which were limed. The relationship between the nutrient treatment and the eventual (1948–1949) species composition of these plots is given in Table 6.1. The data in this table can be interpreted to reveal the relative resource requirements of these species. Because of the number of species and the number of covarying factors, such an analysis is a bit like putting together a large puzzle. Here is how I believe the pieces fall into place.

Consider, first, the unfertilized plots (Plots 3 and 12). These were dominated by *Festuca* (18%), *Leontodon* (14%), and *Agrostis* (10%). The addition of just ammonia (Plot 1) led to a greatly increased dominance by *Agrostis* (Table 6.1), suggesting that *Agrostis* is a poor competitor for nitrogen but a superior competitor for a resource which became more limiting when N was added. Plot 4^2 received both N and P, and had a much lower dominance by *Agrostis* than the plot receiving just N. This suggests that *Agrostis* may be a superior competitor for P, since the addition of P decreases its dominance.

The logic used in drawing this conclusion may seem reversed, so I will illustrate it graphically. Consider two species, one of which is a superior competitor for P but an inferior competitor for N, with the other species being an inferior competitor for P but a superior competitor for N. The discussion above suggests that *Agrostis* is a superior P competitor but an inferior N competitor. The analysis of Figure 59 suggests that the legume, *Lathyrus*, may be a superior competitor for N but an inferior competitor for P. Zero net growth isoclines (ZNGI's) in Figure

TABLE 6.1. Percent abundances of dominant plants in the Rothamsted Park Grass Experiments. Abundance is expressed on a biomass basis. Only those species which are at least 5% of the biomass in a plot are listed. All blanks are for entries less than 5%.

Plot Number	*Treatment*	*Agrostis tennis (vulgaris)*	*Festuca rubra*	*Holcus lanatus*	*Rumex acetosa*	*Anthoxanthum odoratum*	*Leontodon hispidus*	*Alopecurus pratensis*	*Dactylis glomerata*	*Lathyrus pratensis*	*Achillea millifolium*
3, 12	Unfertilized (average of 2 plots)	10	18				14		7		
1	+Nitrogen (N)	75	16								
4[2]	+N, P	36	35	18		10					
18	+N, K, Mg (no P).	77	9		11						
9	+N, P, Mg, K (complete fertilization)	8		91							
10	+N, P, Mg (no K)	52	10	22		10					
11[1]	+N, P, K, Mg (with 50% more N than Plot 9)			99.7							
6, 7, 15	+P, K, Mg (no nitrogen) (Average of 3 plots)		6					11	12	18	9

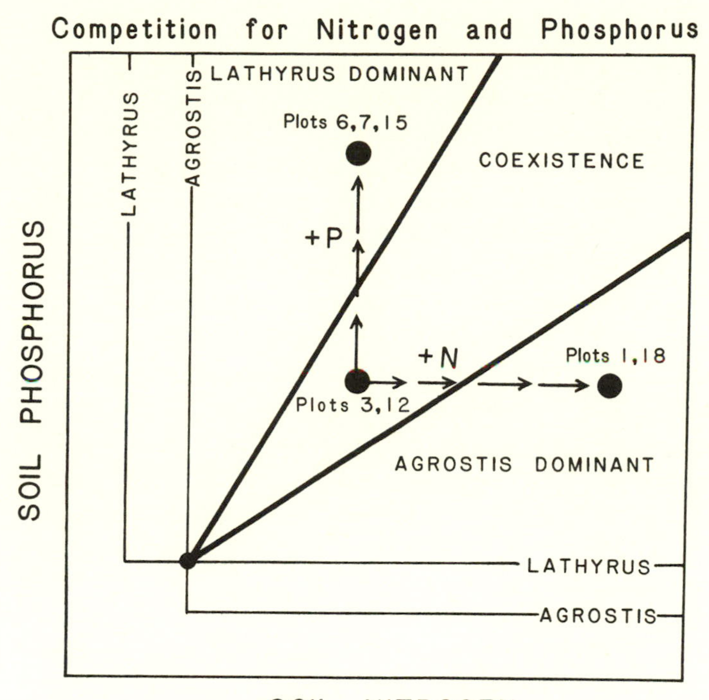

FIGURE 64. The effects of enrichment with P or enrichment with N on the relative abundances of the legume, *Lathyrus pratensis*, and the grass, *Agrostis vulgaris*, are illustrated using a hypothetical ZNGI and consumption vector for each of these species. Enrichment with N favors *Agrostis*, the inferior competitor for N. Similarly, enrichment with P favors *Lathyrus*. *Lathyrus* is favored by P enrichment because P enrichment makes N relatively more scarce, and thus favors the species which is the superior competitor for N, *Lathyrus*.

64 depict such a relationship between *Agrostis* and *Lathyrus*. Consider the point labeled 3, 12 in Figure 64. This point can represent the average supply rate of N and P in Plots 3 and 12, the unfertilized plots. The addition of N (but no P) would move the supply point to the right, as shown. This should lead to a decrease in the relative abundance of *Lathyrus* but an increase in the relative abundance of *Agrostis*. Plot 1 received just N

and Plot 18 received N, K, and Mg, but no P. *Agrostis* became 75% and 77% of the total community in these plots, respectively, a major increase over the 10% that it represented in the unfertilized plots. Consider, also, a case in which P was added but no nitrogen was supplied. The isoclines of Figure 64 suggest that *Lathyrus* should become much more abundant, but that *Agrostis* should decline. Plots 6, 7, and 15 received P, K, and Mg, but no N. *Lathyrus* averaged 18% relative abundance in these plots, compared to 1% in Plots 3 and 12. *Agrostis*, however, averaged 3% in Plots 6, 7, and 15 compared to 10% in the unfertilized plots. As this discussion illustrates, resource competition theory predicts that the species which is the superior competitor for a resource will become less dominant as that resource is added, and that it will become increasingly dominant as the resource for which it is the inferior competitor is added. This occurs because the addition of a resource means that the resource becomes less limiting. Thus, the addition of all resources except that for which a species is the superior competitor would mean that that resource is now in even shorter supply relative to other resources, and that species should become more dominant.

The results of the various Rothamsted fertilizations summarized in Table 6.1 can be similarly interpreted for other species. For instance, the increase in the dominance of *Festuca*, *Holcus*, and *Anthoxanthum* in the plot receiving N and P (Plot 4^2) compared to the plot receiving just N (Plot 1) suggests that these species are superior competitors for a resource or resources which become more limiting when P is added. Plot 18 received all mineral nutrients except P. Plants in Plot 18 are likely limited either by P or by a non-nutrient resource, most probably space or light. As discussed in Chapter 8, it seems unlikely that open space, per se, is a limiting resource in most plant communities. However, the high biomass of the fertilized plot means that very little light will penetrate to the soil surface. This will cause many seedlings and plants of small stature to be light limited. Thus, plants in Plot 18 are likely to be limited either by P or by light. The dominance of *Agrostis* (77%) in Plot 18 and its behavior already discussed above suggest that it is a superior competitor for P. The data for these plots also suggest that *Festuca* is a poor competitor for phosphate.

Plot 9 received complete mineral fertilization, and was dominated by *Holcus* (91%), with *Agrostis* at 8% and all other species rare or absent. With complete mineral fertilization, the resource that should be limiting most of the species present is light. Its dominance in Plot 9 suggests that *Holcus* is the superior competitor for light. Plot 11^1 reinforces this view. This plot received complete mineral fertilization, with 50% more ammonia than Plot 9, and *Holcus* was 99.7% of the biomass on this plot. This later result suggests that the two limiting resources for Plot 9 were light and nitrogen, with *Holcus* and *Agrostis* coexisting in Plot 9 because *Holcus* was limited by nitrogen and *Agrostis* by light. The higher nitrogen of Plot 11^1 meant it was in a region in which light was basically the only limiting resource, and *Holcus* was 99.7% dominant.

The hypothesized interaction between *Holcus* and *Agrostis* in Plots 9 and 11^1 can be illustrated using the graphical approach to resource competition developed in this book. Figure 65 shows ZNGI for these two species growing under conditions of limiting nitrogen and light, but with other resources (phosphate, potassium, etc.) in excess. Unlike nutrients, light is a directional resource, being provided from above. Thus, the main method of light competition (or, if you prefer, interference) is through shading. The ability of one plant to shade another depends on their proximity and relative heights and shapes. For the analysis of Figure 65, it is sufficient to note that light is consumed, and that plants with greater stature generally consume more light. The ZNGI's show *Holcus* being a superior competitor for light and *Agrostis* being a superior competitor for nitrogen. The circle labeled Plot 9 shows the hypothesized spatial heterogeneity in resource supply rates for Plot 9. For such a range of resource

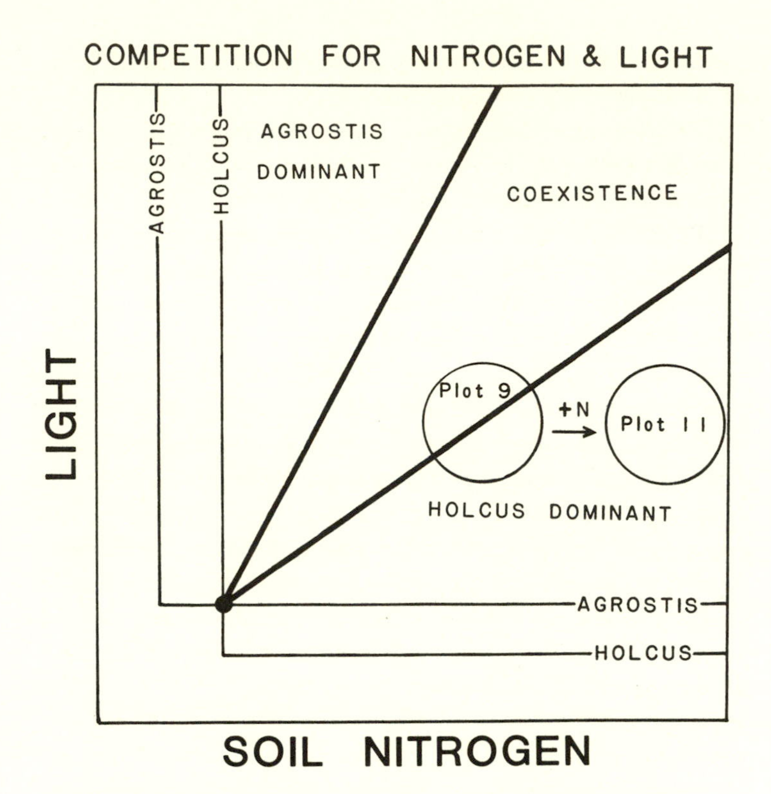

FIGURE 65. The figure shows the hypothesized relative placements of the ZNGI's of *Agrostis vulgaris* and *Holcus lanatus* for limiting nitrogen and light. *Holcus*, which dominates sites with high biomass, is shown as the superior competitor for light (light available for seedlings and shoots), and *Agrostis* is shown as the superior competitor for nitrogen. The shift in their relative abundances from Plot 9 to Plot 11 is consistent with the ZNGI's shown above.

supply rates, both species will coexist, with *Holcus* being much more abundant than *Agrostis*. If Plot 9 were to be fertilized with an even greater amount of nitrogen, its resource supply probability cloud would be moved to the right as shown, giving a plot almost completely dominated by *Holcus*. This corresponds exactly with what happened in Plot 11[1], which differs from Plot 9 only in the additional nitrogen being supplied. Thus, the species composition of these plots seems explicable in terms of the theory developed in this book.

The hypothesis that *Holcus* is a superior competitor for light is further supported by an analysis of the relationship between the percent abundance of *Holcus* in plots and the total biomass of the plots. As the biomass of a plot increases, there will be less light at ground level for the growth of seedlings and shoots. If *Holcus* is a superior competitor for light, it should become increasingly dominant as light becomes more limiting, i.e., as the biomass of a plot increases. Figure 66A shows this. The percent dominance of *Holcus* increases dramatically with biomass in those plots with pH less than 4.0. There is no dependence of its dominance on biomass for plots with pH greater than 4.0. It is interesting to note that the other two species which are superior competitors for light, *Alopecurus* and *Arrhenatherum*, show a similar increase in dominance with increasing biomass, except that their dominance is independent of biomass for pH less than 4.0, and increases with biomass above a pH of 4.0 (Fig. 66B and C). None of the other dominant species in the Rothamsted plots showed such increases in percent abundance with biomass. This further suggests that these three species are specialists on high biomass (low light) sites, and that these three species are separated in their habitat use by pH. Figure 66D illustrates this hypothesis by showing the requirement of each species for light at ground level as it may depend on soil pH. These requirements divide the pH axis into three regions, each of which is predicted to be dominated by a different species. It would be most interesting to determine experimentally the pH and light dependence of the growth rates of these species to discover whether they are consistent with this hypothesis.

The superior competitive ability of these species for limiting light and the dependence of their light requirement on pH may provide an answer to the puzzling coexistence of two species (*Alopecurus* and *Arrhenatherum*) in Plot 14, in which plot all species should have been limited by the same resource, light (see

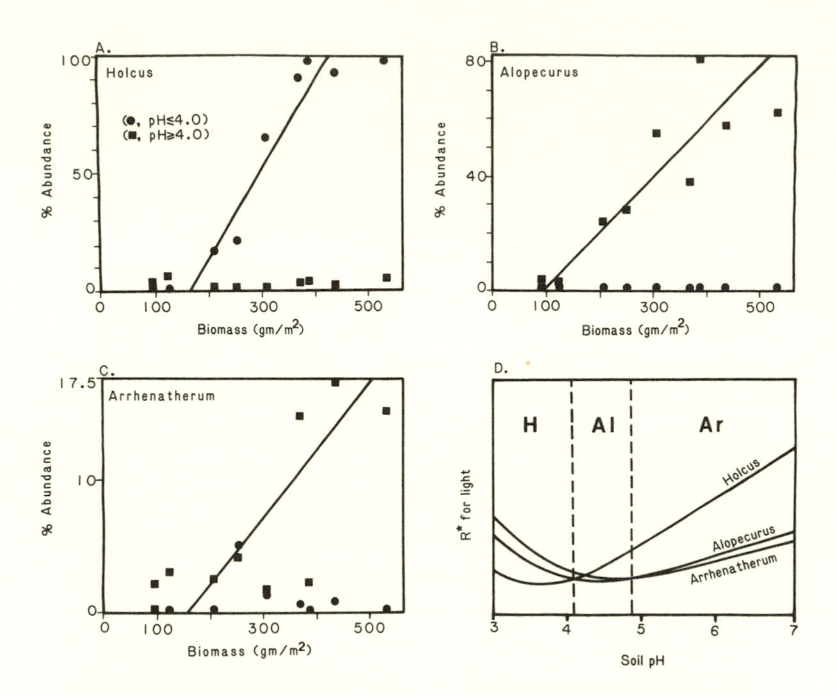

FIGURE 66. *Holcus lanatus*, *Alopecurus pratensis*, and *Arrhenatherum avenaceum* are the three species which dominate plots which receive complete mineral fertilization. The data presented in this figure suggest that these species dominate these plots because the species are superior competitors under conditions of high biomass. For pH less than 4.0, the percent abundance of *Holcus* increases almost linearly with biomass, consistent with the hypothesis that it is a superior competitor for light for this pH range. Similarly, *Alopecurus* increases almost linearly in abundance with biomass for pH greater than 4.0, as does *Arrhenatherum*. The results of this figure, combined with those of Figure 63, strongly suggest that these three species are the superior competitors for light, and that they are separated in their light requirements along a pH gradient. This hypothesis is illustrated in part D, in which the R^* of each species for light is graphed against pH. The regions in which *Holcus*, *Alopecurus*, or *Arrhenatherum* should dominate are labeled H, Al, and Ar, respectively.

Fig. 61). If there is spatial variation in soil pH within Plot 14, it would be possible for these two species to coexist when limited by one resource because each would be a superior competitor in microhabitats with the appropriate pH. Whatever the cause, the long-term, stable coexistence of two species on one limiting resource in Plot 14 is well worth further study.

Additional information on the resource requirements of the species in Table 6.1 comes from Plot 10. Plot 10 received all nutrients except potassium. It was dominated by *Agrostis* (52%), followed by *Holcus* (22%), *Anthoxanthum* (10%), and *Festuca* (10%). Compared to the plots receiving complete mineral fertilizations (Plots 9 and 11[1]), these results suggest that *Holcus* is a poor competitor for potassium, and that *Anthoxanthum* may be the superior competitor for limiting potassium. These results suggest that the structure of the plant community at Rothamsted may be determined by at least four limiting resources (given in the order of approximate importance): nitrogen, light, phosphorus, and potassium.

The relative requirements of the major species for these limiting resources can be used to construct resource isoclines for each species in relation to each of these limiting resources, as was done in Figures 64 and 65 for two-species cases. A complete graphical presentation would require four dimensions, which would be difficult to present on the printed page. However, it is possible to look at various slices through such a four-dimensional structure. Let me illustrate this for nitrogen, light, and phosphorus, considering situations with high potassium. The verbal descriptions of the relative competitive abilities of the species of Table 6.1 for nitrogen, phosphorus, and light were used to construct three such graphs for six major species (*Holcus*, *Agrostis*, *Festuca*, *Dactylis*, *Rumex*, and *Lathyrus*).

Figure 67 shows the hypothesized positions of the ZNGI's of these six species in relation to limiting nitrogen and light. The position of the isoclines was determined using the data of Table 6.1 to surmise the relative competitive abilities of these

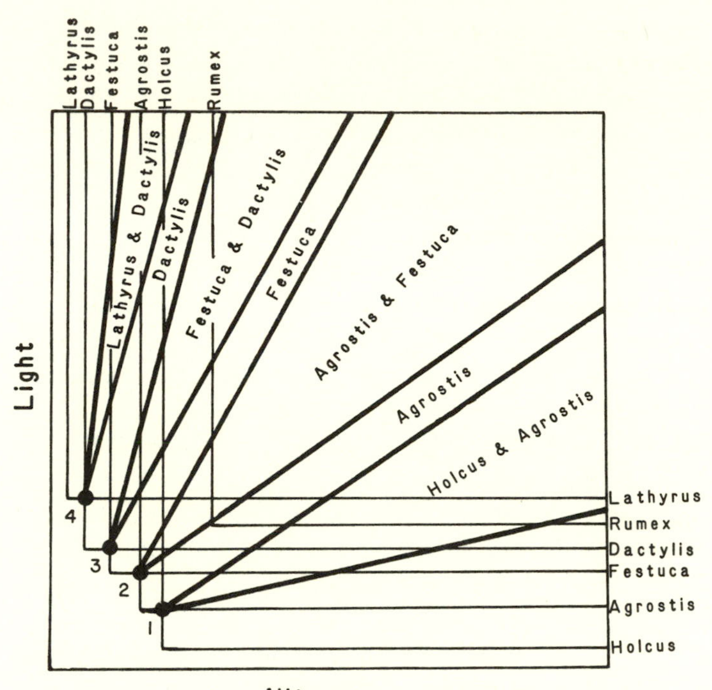

FIGURE 67. Hypothesized relative positions of the ZNGI's of six major plant species in the Rothamsted plots, assuming conditions of high phosphorus and potassium. The ZNGI's for limiting light and nitrogen were derived using the data of Table 6.1 and the method of logic illustrated in Figures 64 and 65. Because they are based on the data of Table 6.1, these hypothesized ZNGI's should apply to low pH plant communities. The major difference for higher pH communities would be the substitution of *Alopecurus* and then *Arrhenatherum* for *Holcus*.

species for either limiting nitrogen or light. These isoclines show *Holcus* as the superior competitor for light, followed by *Agrostis*, *Festuca*, *Dactylis*, *Rumex*, and *Lathyrus*. The legume, *Lathyrus*, is shown as the superior competitor for nitrogen, followed by *Dactylis*, *Festuca*, *Agrostis*, *Holcus*, and *Rumex*. These ZNGI's lead to four two-species equilibrium points, shown with large dots, and to the labeled regions of coexistence of pairs of species. As a starting point in understanding the derivation of these ZNGI's,

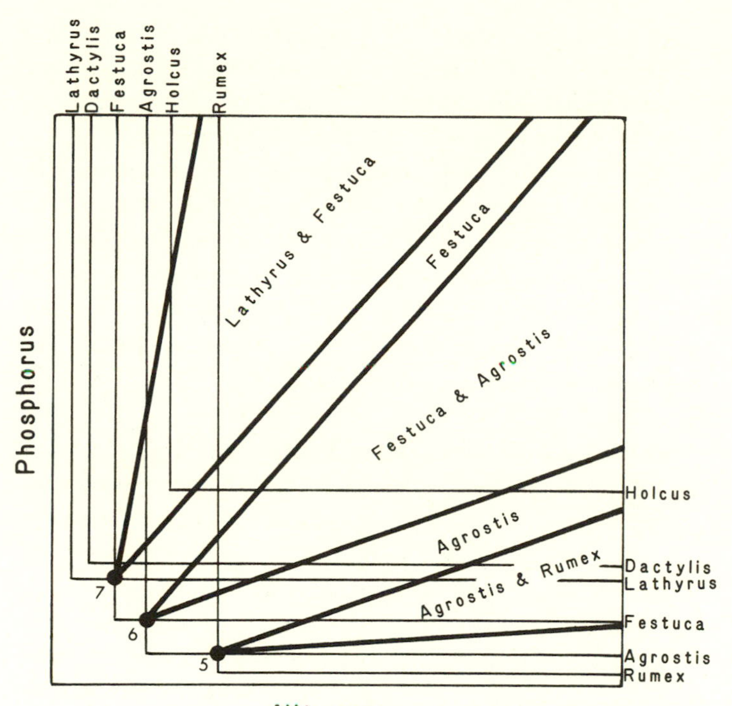

FIGURE 68. Hypothesized relative positions of the ZNGI's of six Rothamsted plants when competing for nitrogen and phosphorus under conditions of high levels of light and potassium. These were derived using the data of Table 6.1.

consider *Holcus* and *Agrostis*. As already discussed, these species coexist in Plot 9 because *Holcus* is limited by nitrogen and *Agrostis* by light. At the *Holcus-Agrostis* equilibrium point of Figure 67 (this equilibrium point is numbered 1), this condition is met. Similarly, *Dactylis* is limited by light and *Festuca* is limited by nitrogen at equilibrium point 3 of Figure 67. *Rumex* is shown as an inferior competitor for both nitrogen and light, and should not occur in plots in which these are the only limiting resources.

The hypothesized ZNGI's for conditions of limiting phosphorus and nitrogen are shown in Figure 68. The competitive ability shown for limiting nitrogen is exactly the same as that of Figure 67. The additional information in this figure is the relative competitive abilities of these species for phosphorus. From the data of Table 6.1, *Rumex* is hypothesized to be the superior competitor for phosphorus, followed by *Agrostis*, *Festuca*, *Lathyrus*, *Dactylis*, and *Holcus*. These ZNGI's lead to three two-species equilibrium points, labeled 5, 6, and 7. These show that *Rumex* should dominate plots low in phosphorus but rich in all other resources, and that *Lathyrus* should dominate plots low in nitrogen but rich in all other resources. *Festuca* and *Agrostis* should dominate habitats with intermediate availabilities of N and P.

Figure 69 shows the isoclines for limiting phosphorus and light but excess nitrogen. It does not contain any new information, since it uses the same position for the phosphorus segment of the ZNGI as in Figure 68 and the same position for light as in Figure 67. However, it is useful, since it allows easy prediction of the effects of the assumptions of Figures 67 and 68 on competition between these species under conditions of limiting light and phosphorus. There are only two two-species equilibrium points. Point 8 states that *Rumex* and *Agrostis* should coexist for conditions of low phosphorus and moderately low light, when all other resources are in high abundance. Equilibrium point 9 means that *Holcus* and *Agrostis* should coexist for conditions of very limiting light and moderately low availabilities of phosphorus, when other resources are not limiting.

Figure 68 may be usefully compared with Figure 59. Figure 68 shows the hypothesized positions of the ZNGI's of six species for nitrogen and phosphorus, based on the data of Table 6.1. Figure 59 shows the dependence of the relative abundances of several of these species on the ratio of soil N to soil P. Because these are two different ways to describe the results of the experiments at Rothamsted, these are not independent pieces of information. The two figures were derived independently from different data sets, and so similarities between these different

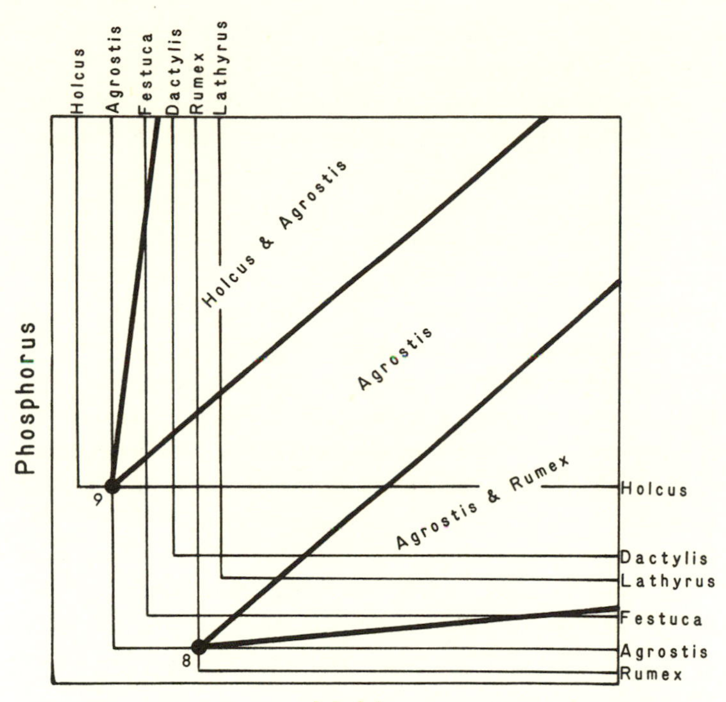

FIGURE 69. Hypothesized relative positions of the ZNGI's of six Rothamsted plants for limiting light and phosphorus, under conditions of high levels of nitrogen and potassium. These ZNGI's were derived using the data of Table 6.1. The ZNGI's in this figure are consistent with those of the preceding two figures, and could be derived directly from those figures.

methods of analysis may be considered indicative of consistent patterns in the Rothamsted experiments. The soil pH of the two data sets are different, with Figure 59 summarizing higher pH sites and Figure 68 summarizing lower pH sites. Figure 59I shows that several species are separated along a N:P ratio gradient, with the legume *Lathyrus* most abundant at low ratios (nitrogen limitation), followed by the grasses *Arrhenatherum*, *Festuca*, *Poa*, *Alopecurus*, and *Agrostis*, followed by the herb

Leontodon, which is most dominant on the high N:P ratios indicative of phosphorus limitation. The ZNGI's of Figure 68 suggest that the legume *Lathyrus* should dominate the most nitrogen-limited habitats, followed by the grasses *Festuca* and *Agrostis* and then the herb *Rumex* in increasingly nitrogen-rich and phosphorus-poor habitats. The order of species along a N:P gradient for the species that overlap in the two analyses (i.e., that occur over a broad pH range) is exactly the same. Interestingly, the two species which did not show any clear dependence of their abundance on N:P ratios in Figure 59, *Dactylis* and *Holcus*, are predicted by the isocline approach of Figure 68 to be inferior competitors for both nitrogen and phosphorus relative to the other species. Thus, no dependence of their relative abundance on N:P ratios is expected. The similarity of these two different methods of analysis is strong suggestive evidence of the role of nitrogen and phosphorus competition in determining the species composition of the Rothamsted plots. An intriguing correlate of these studies is Rorison's (1968) finding that *Rumex acetosa* had the lowest phosphate requirement of four species that he studied. (See Fig. 1B.)

If the ZNGI's of Figures 67, 68, and 69 represent the requirements of these six species for these three commonly limiting resources, it should be possible to superimpose on each figure the resource supply probability cloud for each of the experimental plots. Figure 70 shows such resource supply probability clouds as circles, with each cloud numbered to correspond with a given plot at Rothamsted (see Table 6.1). For ease of interpretation, the isoclines of species which are not part of a two-species equilibrium point are eliminated. Figure 70 shows the hypothesized position of the resource supply probability distributions for the unfertilized control plots (3 and 12), for the plots which received no nitrogen (6, 7, and 15), and for Plots 1, 4^2, 9, and 11^1, all of which received nitrogen. It graphically illustrates the decreased abundance of *Agrostis* and the increased

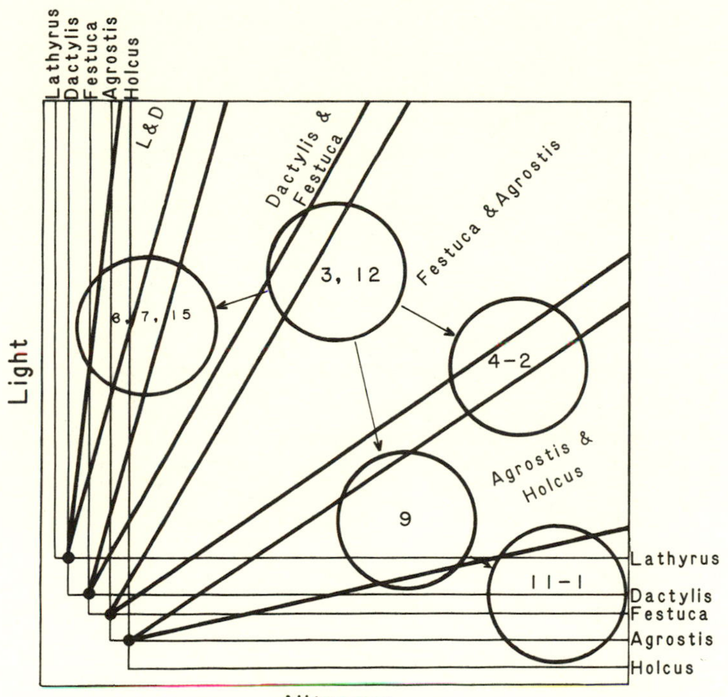

FIGURE 70. Superimposed on the ZNGI's from Figure 67 are the approximate relative positions of the resource supply probability clouds for several of the Rothamsted plots. Because fertilization caused plots to have greater biomass than the unfertilized plots, light (available at the soil surface) decreased in the fertilized plots relative to Plots 3 and 12.

abundance of *Holcus* on Plot 11[1] compared to Plot 9. Comparison of the positions of these hypothesized resource supply distributions with the data of Table 6.1 indicates that the proper species are dominant in each of the plots. It also illustrates how the addition of N, N and P, and everything but N compared to the controls may have led to changes in the species composition of these plots.

Similar relationships are shown in Figure 71 for the effects of N and P on species composition. Again, the unfertilized plots

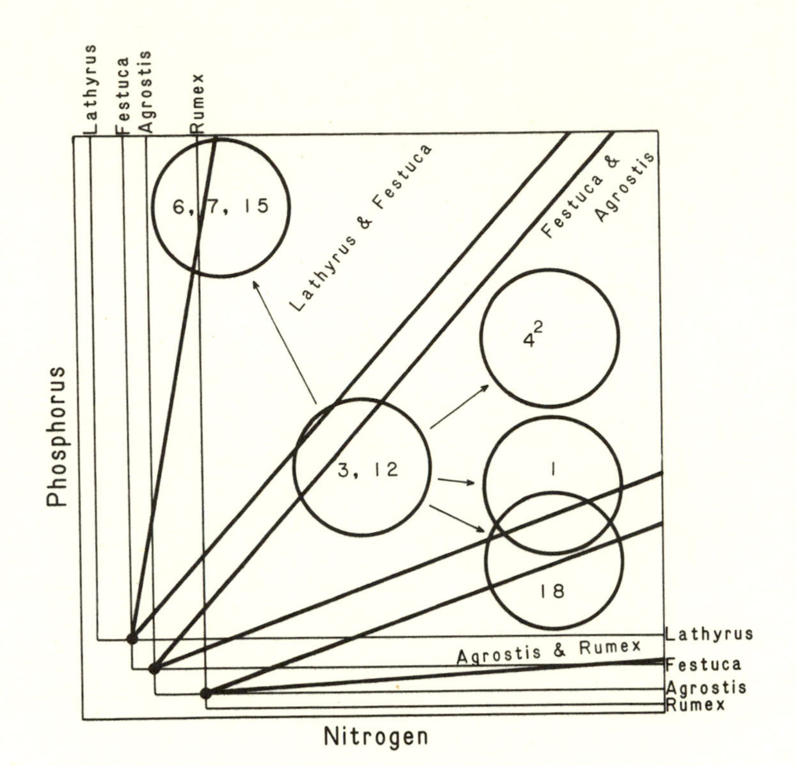

FIGURE 71. Superimposed on the ZNGI's of Figure 68 are the approximate positions of the resource supply probability clouds for some of the Rothamsted plots. The decrease in nitrogen shown for Plots 6, 7, and 15 compared to unfertilized Plots 3 and 12 is consistent with the chemistry data reported by Warren and Johnston (1963).

are shown, as are the effects of addition of all nutrients except P (Plot 18), just N (Plot 1), and N and P (Plot 4^2) on species composition. Figure 72 shows the hypothesized position of Plots 9, 11^1, and 18 in relation to the supply rates of light and phosphorus. For all of these cases, the relative dominance of species is consistent with the data of Table 6.1. This clearly demonstrates that the major results of all the manipulations at Rothamsted can be interpreted in a manner which is consistent

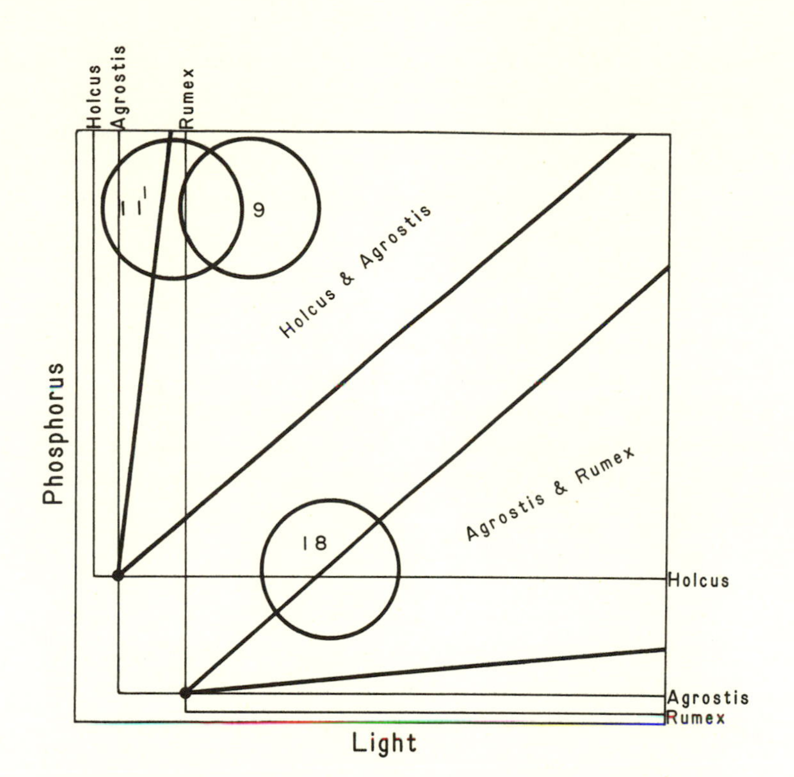

FIGURE 72. Superimposed on the ZNGI's of Figure 69 are the hypothesized positions of the resource supply probability clouds of some of the Rothamsted plots, for plots in which light and phosphorus are limiting resources.

with a theory of resource competition in spatially heterogeneous environments.

This exploration of the results of the Rothamsted experiments is strong suggestive evidence that a resource-based approach such as that developed in this book may be able to predict the patterns of species dominance in natural plant communities. The different patterns of fertilization at Rothamsted have led to communities with very different species compositions. It is possible to use this information to generate resource isoclines, and thus to describe the hypothesized competitive abilities of these species for the major limiting resources. There are two ways to test these predictions. The first is through careful study of the resource requirements of these species under controlled conditions, including experiments on the nutrient and light requirements of the species grown singly, and studies of pairwise and multispecies competition for the resources which are limiting in the natural habitat. The second is through correlational studies in the fields adjoining the experimental plots at Rothamsted, to determine the relationships between the supply rates of these resources and the dominance of the species in various microsites. Both of these tests, together, could provide the most rigorous test of any community-level hypothesis ever performed in ecology.

SUMMARY

The physiological, correlational, and experimental data summarized in this chapter are generally consistent with the hypothesis that resource competition is an important component of the processes that determine the species composition of natural plant communities. There are striking consistencies between the optimal resource ratio of a species as determined in the laboratory and the ratio at which the species dominates natural communities. Correlational and experimental evidence for nitrogen-fixing organisms (legumes and blue-green algae) strongly supports the hypothesis that they dominate habitats with a low ratio of nitrogen to other limiting resources. The resource isocline approach seems capable of describing the observed changes in the species composition of the Rothamsted plots. However, none of the studies are definitive. All are subject to alternative interpretations. This will continue until there has been a sufficiently thorough study of a few selected plant communities that all of the ramifications of a theory of resource competition may be tested in them. To test this theory thoroughly, it will be necessary to know the resource requirements and competitive interactions of the dominant species under controlled conditions, the correlations between the distributions of these species in the field and the distributions of limiting resources, and the effects of various enrichments on the species composition of natural communities. However, the work reviewed in this chapter strongly suggests that resources are a major factor influencing the diversity and species composition of plant communities.

CHAPTER SEVEN

A Comparison with Classical Competition Theory

The preceding six chapters have presented a mechanistic, resource-based approach to competition, and illustrated its possible applicability to natural communities. The theory I have developed is quite different, in spirit, from the classical Lotka-Volterra competition equations which have been the major descriptor of competition in the ecological literature since the 1920's (Lotka, 1924; Volterra, 1931; Gause, 1934).

The resource-based theory attempts to predict the outcome of competition by using information on the resource requirements of the various competing species, whereas the Lotka-Volterra approach describes the interactions between species in terms of summary variables, the competition coefficients. Despite these differences in purpose, there are interesting analogies between the two approaches. May (1973) has shown that the Lotka-Volterra model may be considered a near-equilibrium approximation to many other models of competition.

Because of the prevalence of the Lotka-Volterra model in textbooks, in current ecological research; and in the thought processes of ecologists, I would like to explore the equilibrium relationships between the Lotka-Volterra approach and some resource-based approaches to competition. I believe that this is especially important because of some of the generalizations which have been made concerning resource competition and the Lotka-Volterra competition coefficient. Should the competition coefficient, alpha, be constant along a resource gradient, should it change smoothly as some have suggested, or should it change in a stepwise fashion? How does the competition coefficient depend on the type of resource for which competition occurs, on the supply processes of the resources, and on the consumption characteristics of the species?

In order to answer these questions, it is necessary to formulate an explicit model of resource competition. Once this is done, the equilibrium outcome of competition can be equally well explained using either the graphical approach of the preceding chapters or the classical Lotka-Volterra approach. The results of such analyses suggest that many of the studies of species packing and limiting similarity (e.g., MacArthur and Levins, 1967; May and MacArthur, 1972; May, 1973) should be expanded to determine how dependent the results are on the type of resource for which competition occurs.

The Lotka-Volterra equations for competition between two species may be written as

$$\begin{aligned} dN_1/dt &= r_1N_1(K_1 - N_1 - \alpha N_2)/K_1, \\ dN_2/dt &= r_2N_2(K_2 - N_2 - \beta N_1)/K_2, \end{aligned} \tag{16}$$

where N_1 and N_2 are the population densities of species 1 and 2, r_1 and r_2 are the maximal per capita rates of increase of species 1 and 2, K_1 and K_2 are the carrying capacities of species 1 and 2, and α and β are the competition coefficients (see any introductory ecology textbook). The competition coefficient α measures how much each individual of species 2 depresses the potential carrying capacity of species 1. Similarly, the competition coefficient β measures how much each individual of species 1 reduces the potential carrying capacity of species 2. The meaning of α and β may be more easily seen by setting $dN_1/dt = dN_2/dt = 0$. This gives two equations for population density growth isoclines for N_1 and N_2:

$$\begin{aligned} &dN_1/dt = 0 \quad \text{means that} \quad N_1 = K_1 - \alpha N_2, \qquad \text{and} \\ &dN_2/dt = 0 \quad \text{means that} \quad N_2 = K_2 - \beta N_1. \end{aligned} \tag{17}$$

In order to obtain the analogs of α and β from a given model of resource competition, it is necessary to solve the resource model for equilibrium (i.e., for $dN_1/dt = dN_2/dt = dR_1/dt = dR_2/dt = 0$) and then rework the equations until they are in the same form as shown above. In doing this reworking, it is necessary to solve for the carrying capacity of each species, i.e., the population density that each species would reach in the absence of a competitor. The results of such calculations for cases of competition for essential, perfectly substitutable, and switching resources are discussed and graphically illustrated in the following pages. It is possible to perform comparable analyses for any other resource class. However, I want to note that I do not present these calculations because I believe that the usefulness of a resource-based approach to competition comes from its ability to predict the parameters of the Lotka-Volterra competition equations. Rather, I present them to illustrate the interrelationships of several resource models and classical theory.

Perfectly Substitutable Resources

Consider, first, a case of two species competing for two perfectly substitutable resources. One such case is illustrated in Figure 73. As drawn in Figure 73, each species consumes more of the resource which more limits its own growth. Thus the equilibrium point is locally stable. As before, the ZNGI's and consumption vectors of Figure 73 define regions in which neither species can survive, in which only one species can survive, in which both can survive but one or the other species will be dominant, and a region in which both species can stably coexist. In determining the competition coefficients, I will only consider supply points which would lead to a stable population of either species if the species were growing by itself. As the calculations in the Appendix demonstrate, α and β are constant, independent of the resource supply point, for all such resource supply points. Using the relationships derived in the

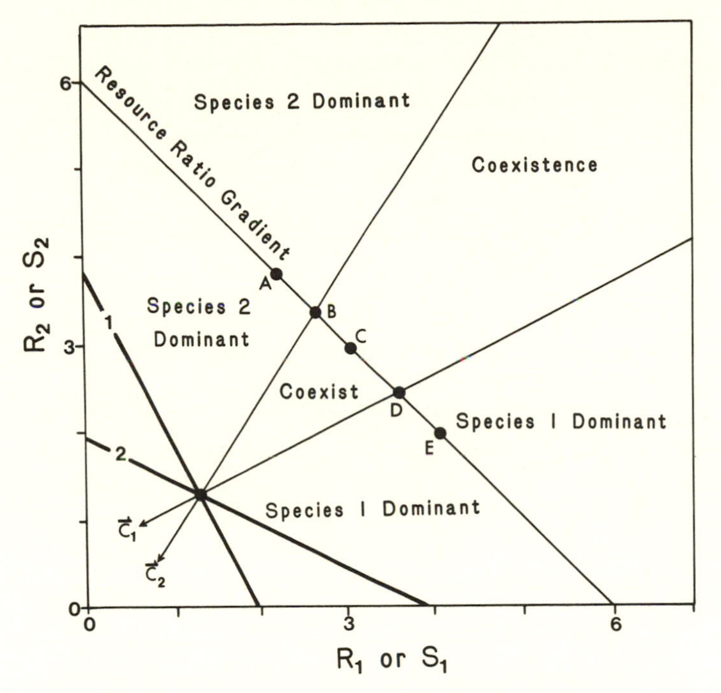

FIGURE 73. Competition between two species for two perfectly substitutable resources. The ZNGI's and consumption vectors of species 1 and 2 are labeled 1 and 2. The two-species equilibrium point is indicated with a dot. The points labeled *A* to *E* on the Resource Ratio Gradient are discussed in the text and are used as reference points for the Lotka-Volterra analysis illustrated in Figure 74.

Appendix, the values of α, β, K_1, and K_2 for the case illustrated in Figure 73 are presented in Figure 74, along with the usual Lotka-Volterra competition diagrams.

Figure 74 emphasizes five cases, each from a different point along the resource ratio gradient of Figure 73. Points along this gradient range from supply of only R_2, to intermediate rates of supply of both resources, to supply of only R_1. Point *A* (see Fig. 73) is a resource supply point in the region in which species 2 should competitively displace species 1. The Lotka-Volterra $N_1 - N_2$ phase plane diagram for point *A* is shown

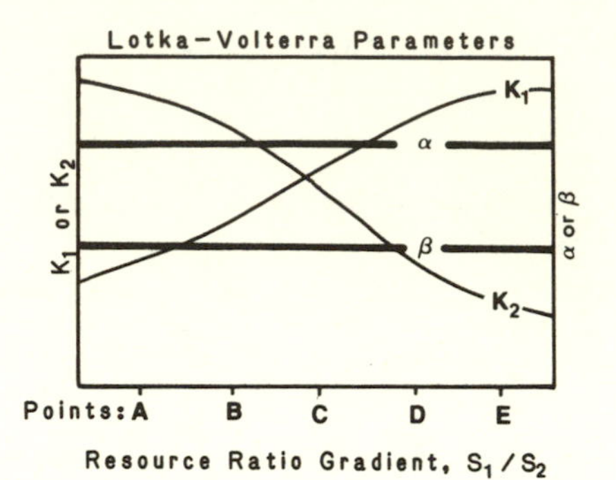

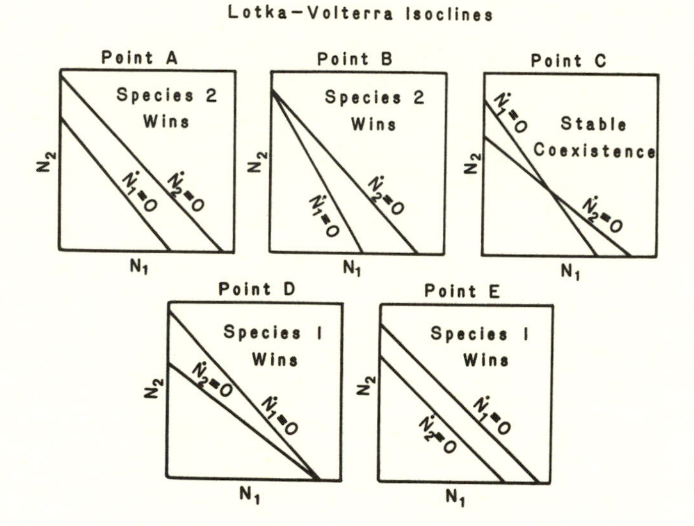

FIGURE 74. The upper part of this graph shows the predicted values of Lotka-Volterra competition equation parameters (α, β, K_1, and K_2) for various points along the Resource Ratio Gradient of Figure 73. Note that the competition coefficients are constant along this gradient, and that the carrying capacities change. The Lotka-Volterra isoclines of the lower part of this figure were constructed using these values of the competition coefficients and the carrying capacities. According to the Lotka-Volterra isoclines, species 2 should competitively displace species 1 for point *A* on the Resource Ratio Gradient. This prediction is identical to that of the graphical approach to resource competition illustrated in Figure 73. Similarly, the outcomes of competition predicted by the two methods of analysis for other points along the Resource Ratio Gradient are identical, thus demonstrating the interchangeability of the two approaches to the equilibrium outcome of competition. Note that $\dot{N}_i$ is a shorthand notation for dN_i/dt.

in the lower section of Figure 74. According to Lotka-Volterra theory, species 2 should competitively displace species 1 for resource supply point *A*. Point *B* is exactly on the border between coexistence and dominance by species 2. The graphical approach to resource competition theory presented earlier in this book predicts that species 2 should approach its carrying capacity, and species 1 should have an equilibrium population density of zero at resource supply point *B*. The Lotka-Volterra competition coefficients and carrying capacities calculated for point *B* make this same prediction, as indicated by the isoclines for this case in Figure 73. Both species should stably coexist for point *C*. The Lotka-Volterra parameters calculated for supply point *C* lead to a phase plane diagram which describes stable coexistence (Fig. 74). For point *D*, the Lotka-Volterra isoclines show that species 1 will win, with species 2 attaining an equilibrium population density of zero. Similarly, the Lotka-Volterra isoclines predict that species 1 should competitively displace species 2 for supply point *E*.

This analysis shows that there is a true, equilibrium analogy between the Lotka-Volterra competition equations and the graphical model of competition for substitutable resources. The outcome of competition for any given resource supply point is identical whether the outcome is predicted using the graphical approach to resource competition developed in this book or the $N_1 - N_2$ phase plane method associated with the Lotka-Volterra theory.

As shown in Figure 74, the competition coefficients of the Lotka-Volterra equations do not change along a resource ratio gradient for perfectly substitutable resources supplied in an "equable" mode (see Eq. 10 in Chapter 4). The different predicted outcomes of competition result from changes in the carrying capacities of the species along the resource ratio gradient (Fig. 74). This case of competition for perfectly substitutable resources may be contrasted with the case explored by MacArthur (1972). The model used above is identical to

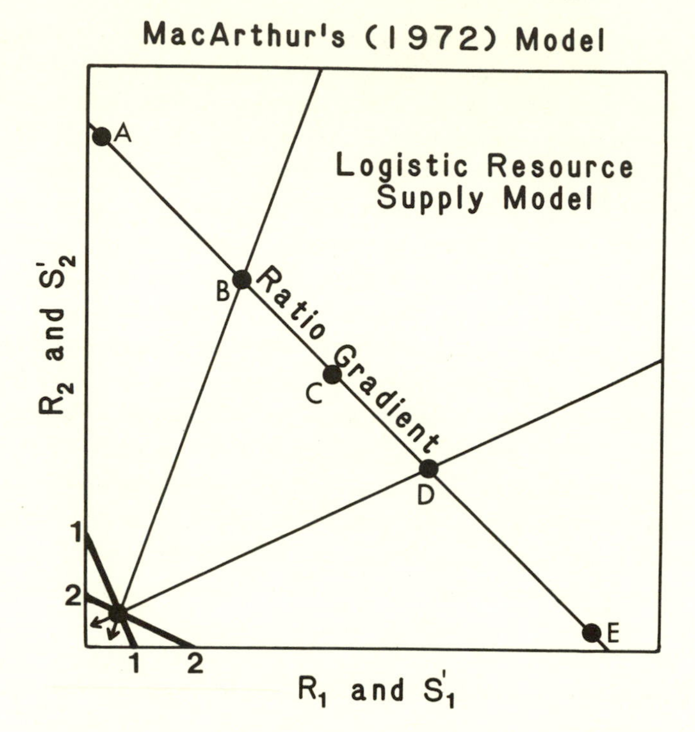

FIGURE 75. A case of competition for two perfectly substitutable resources in which resources are supplied according to the logistic model of MacArthur. Because this model predicts different resource supply vectors for a given "resource supply point" than does the equable model, the predicted outcomes of competition are not indicated on the graph above.

that of MacArthur (1972) except that I have assumed that resources are supplied in an equable mode whereas MacArthur assumed that resources were supplied in a "logistic" manner. Thus MacArthur's model had $dR_i/dt = a(S_i' - R_i)/S_i'$. This difference makes the α and β that MacArthur calculated dependent on S_1' and S_2', where S_1' and S_2' are the logistic "carrying capacities" of resources 1 and 2, respectively. As this example will illustrate, (S_1', S_2') is not the same as the resource supply point, for a different model is used to determine the rate of resource supply for a given deviation away from S'_1 and S'_2 than for a deviation away from S_1 and S_2. Figure 75 shows the ZNGI's and consumption vectors for one case of MacArthur's model, and Figure 76A shows the values for α, β, K_1, and K_2 calculated using MacArthur's formulae. These lead to the $N_1 - N_2$ phase plane diagrams of Figure 76B. Species 2 wins at point A, the species coexist for points B, C, and D, and species 1 wins for point E. That points A to E are not resource supply points is clearly illustrated by the coexistence of species 1 and 2 at point B. Point B is not in the region in which a resource supply point must fall for the two species to coexist stably. A resource supply point at point B would lead to dominance by species 2, with species 1 being competitively displaced. However, the two species coexist when (S'_1, S'_2) is at point B. This occurs because, using MacArthur's model of logistic resource supply, the (S'_1, S'_2) at point B gives a resource supply vector, $\vec{U}$, from the two-species equilibrium point which falls within the region of coexistence. Similarly, a resource supply point at D would lead to dominance by species 1, with species 2 being competitively displaced. However, (S'_1, S'_2) at this point leads to coexistence, again because the resource supply vector from the two-species equilibrium point falls within the region of coexistence when resources are supplied logistically.

This example illustrates the broad applicability of the simple criteria presented in Chapter 4 for determining the equilibrium outcome of resource competition. Two species may coexist when the resource supply vector, $\vec{U}$, falls between the region defined by the consumption vectors of the two species. The actual function defining the resource supply process is relatively unimportant. However, it is only for the idealized case of equable resources that the concept of a resource supply point may be used. With "equable" resources, the resource supply vector always points from a given point straight at the resource supply point. Thus, it is possible to predict the outcome of competition for equable resources very easily using graphical

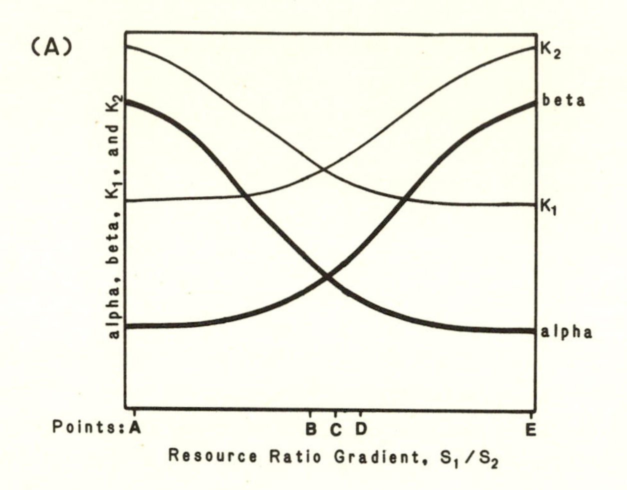

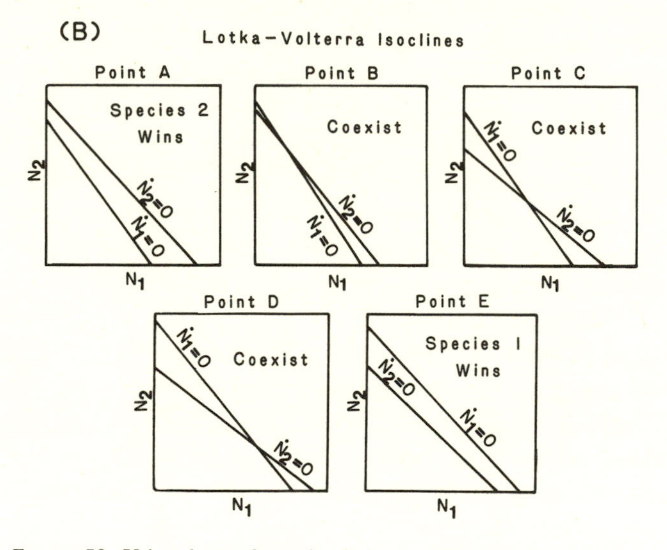

FIGURE 76. Using the mathematics derived by MacArthur (1972), this figure illustrates the dependence of the competition coefficients and the carrying capacities on position along the Resource Ratio Gradient. Part B shows that the two species coexist for points *B*, *C*, and *D* along this ratio gradient, a result consistent with the resource supply vectors associated with each of these logistic resource supply points. See text for more details.

methods, as has already been discussed. It is also possible to approximate other more complex resource supply processes, at equilibrium, with the equable model. This may be done by calculating the resource supply point associated with any given observed resource supply vector.

Essential Resources

The relationship between the Lotka-Volterra model and a resource-based model of competition for essential resources supplied in an equable mode is illustrated in Figure 77.

For essential resources, the values for α and β are not constant along a resource ratio gradient, but depend on which resource limits the growth of each species. Similarly, the carrying capacities of each species depend on which resource is limiting the growth of the species. There are three critical regions of the $R_1 - R_2$ plane. In region A of Figure 77 both species are limited by R_1. Here the equilibrium values of α, β, K_1, and K_2 are

$$\begin{aligned} \alpha &= \frac{c_{21}}{c_{11}}, \\ \beta &= \frac{1}{\alpha} = \frac{c_{11}}{c_{21}}, \\ K_1 &= \frac{D}{c_{11}}(S_1 - R^*_{11}), \qquad \text{and} \\ K_2 &= \frac{D}{c_{21}}(S_1 - R^*_{21}), \end{aligned} \tag{20}$$

where c_{ij} is the amount of R_j consumed by species i, R^*_{ij} is the amount of R_j required for species i to maintain a stable population in the absence of competition, i.e., its ZNGI amount of R_j, and D is the per capita mortality rate. In region B, in which each species is limited by a different resource, the values of

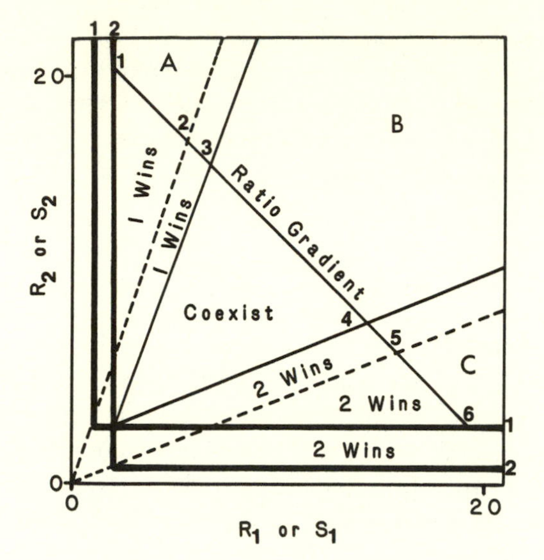

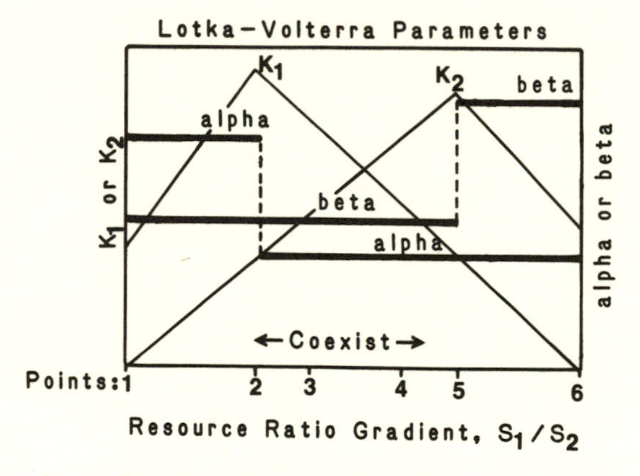

FIGURE 77. A. The ZNGI's and regions of dominance and coexistence of species 1 and 2, which are competing for essential resources.

B. The ZNGI's and consumption vectors of part A, above, lead to the carrying capacities and competitition coefficients shown.

these parameters are

$$\begin{aligned} \alpha &= \frac{c_{22}}{c_{12}}, \\ \beta &= \frac{c_{11}}{c_{21}}, \\ K_1 &= \frac{D}{c_{12}}(S_2 - R_{12}^*), \qquad \text{and} \\ K_2 &= \frac{D}{c_{21}}(S_1 - R_{21}^*). \end{aligned} \tag{21}$$

Both species are limited by R_2 in region C, giving Lotka-Volterra parameters of

$$\begin{aligned} \alpha &= \frac{c_{22}}{c_{12}}, \\ \beta &= 1/\alpha = \frac{c_{12}}{c_{22}}, \\ K_1 &= \frac{D}{c_{12}}(S_2 - R_{12}^*), \qquad \text{and} \\ K_2 &= \frac{D}{c_{22}}(S_2 - R_{11}^*). \end{aligned} \tag{22}$$

Using these values for the Lotka-Volterra parameters for any given resource supply point, it is possible to calculate the outcome of competition predicted by the Lotka-Volterra model. Again, the outcome predicted using the Lotka-Volterra parameters is identical to that predicted by the graphical model of resource competition. The values of α, β, K_1, and K_2 along the resource ratio gradient of Figure 77A are shown in Figure 77B, and the region of coexistence is indicated. Note that, for essential resources and equable resource supply, the values of the competition coefficients do not change smoothly along a resource ratio gradient, but abruptly change value as a species

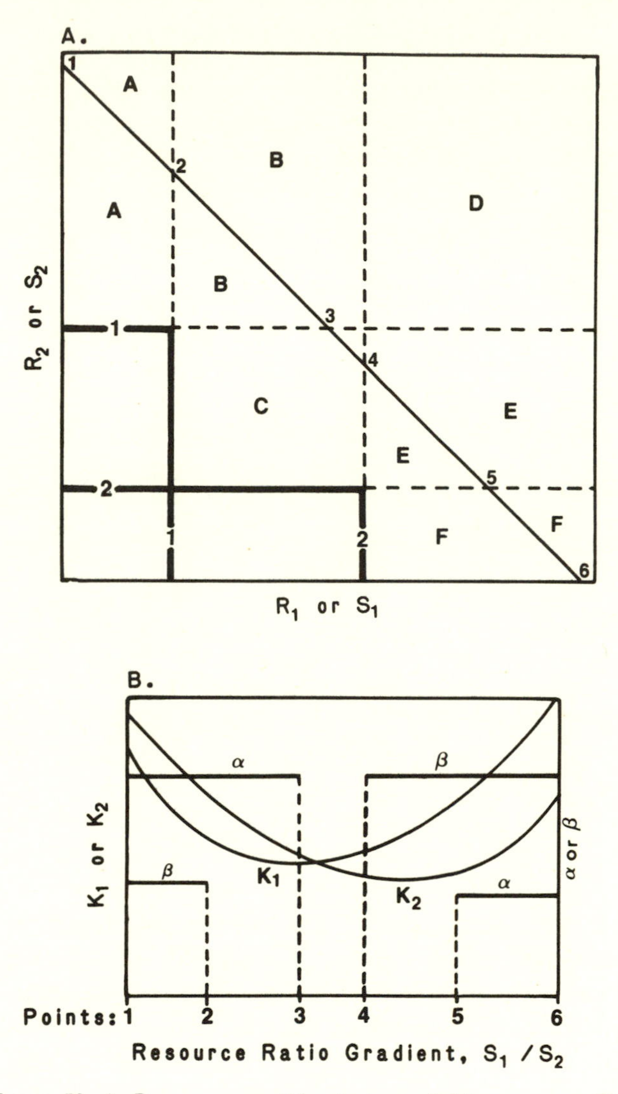

FIGURE 78. A. Resource competition for two switching resources. The predicted values of the competition coefficients change from one region to the next. For regions *A* to *F* in this figure, the predicted values of the competition coefficients are as follows, where α is given first and β second: region A—c_{22}/c_{12} and c_{12}/c_{22}; region *B*—c_{22}/c_{12} and 0; region *C*—0 and 0; region D—c_{22}/c_{12} and c_{11}/c_{21}; region *E*—0 and c_{11}/c_{21}; region *F*—c_{21}/c_{11} and c_{11}/c_{21}.

B. The dependence of these parameters on position along a Resource Ratio Gradient is illustrated.

goes from being limited by one to being limited by the other resource.

Switching Resources

The final analogy between resource models and Lotka-Volterra models which I will discuss is for switching resources. The resource requirements of the two species of Figure 78A define six regions of resource supply in which α, β, K_1, and K_2 take on different values. These regions, and the associated values of the competition coefficients are given in Figure 78A. The values for these parameters along the resource ratio gradient shown are given in Figure 78B. The outcome of competition calculated using the Lotka-Volterra coefficients is identical to that calculated using the graphical approach to resource competition.

SUMMARY

These comparisons of the Lotka-Volterra and the resource approaches to competition demonstrate that the two models may be used interchangeably at equilibrium, as noted by May (1973). However, they also demonstrate that it is necessary to know the explicit, mechanistic model of resource competition in order to determine how the parameters of the Lotka-Volterra equations will change along a resource ratio gradient. The two examples of competition for perfectly substitutable resources, one with resources supplied in an equable mode and one with resources supplied logistically, make this point very well. When perfectly substitutable resources are supplied in a equable mode, α and β are constant along a resource ratio gradient. The changes in dominance patterns along the gradient are caused by changes in the carrying capacities of the two species. However, when resources are supplied logistically, both the competition coefficients and the carrying capacities of the species change along a resource ratio gradient. For perfectly substitutable resources supplied logistically, the competition coefficients change smoothly along the resource gradient, whereas for either essential or switching resources supplied equably, the competition coefficients change abruptly at different points along the resource gradient. These patterns suggest that it may be dangerous to use the Lotka-Volterra approach to model cases of competition along a resource gradient because it is difficult to know which mechanisms of competition are reflected in a given pattern of change of the competition coefficients and carrying capacities along the gradient.

The work of MacArthur and Levins (1967), May and MacArthur (1972), May (1973), Abrams (1975), Yoshiyama and Roughgarden (1977), Turelli (1978), and others has all contributed to increasing our understanding of the limits to similarity of competing species. However, most of these studies have not explicitly considered the mechanisms of competition and have assumed competition for perfectly substitutable resources. Future work may benefit from explicitly exploring competition for various types of resources to ascertain to what extent the limits to similarity depend on the types of resources for which competition occurs, the mode of resource supply, and species' consumption characteristics.

CHAPTER EIGHT

Space as a Resource, Disturbance, and Community Structure

The importance of periodic disturbance to the maintenance of local species diversity has been recognized since the experiments of Darwin (1859). In some of the work that has been done on the effects of disturbance on the diversity and species composition of communities of sessile organisms, a predator or herbivore has been the cause of the disturbance (e.g., Paine, 1966, 1969; Lubchenco, 1978; Harper, 1969; Platt and Weis, 1977), whereas in other studies, non-biological processes, such as wave-caused disturbances (e.g., Dayton, 1971), wind-caused tree falls (Horn, 1971; Fox, 1977; Denslow, 1980; Sprugel and Bormann, 1981), or landslides (Garwood, Janos, and Brokaw, 1979) are the mechanisms opening up new sites for colonization. Whatever the cause, disturbance in communities of sessile organisms provides sites in which new individuals can become established. Such sites are essential for many immobile organisms because they provide a physically open area for colonization as well as other resources associated with an opening. Thus, disturbance can be thought of as a process that provides one or more required resources.

There are two different connotations which have become associated with the concept of space as a resource. In one case, physically open space, itself, is considered a resource without respect to any other variables. In the other case, an "open site" is considered a composite resource which summarizes the availabilities of numerous distinct resources associated with the open site. For a plant community, an open site refers to a site which is open to sunlight and which lacks other plants. The absence of other plants means that soil nutrients and water will be available for a colonist, and that there will be no other plants interfering with the plant by shading it from sunlight. It also means that there is a physically open area in which a plant may become established. For communities of sessile aquatic animals, an open site has qualities such as orientation with respect to the rate and direction of water flow, and food availability associated with such water flow, as well as being a physically open area on which an individual may settle. I would like to maintain a distinction between these two views. Throughout this discussion, I will refer to the composite, summary resource as "open sites," and will refer to a physically open space on which an organism can become established as "space" or "open space." Thus, space as a resource refers to a physically open space on which an individual can potentially become established, but it in no way implies any other qualities about the space. An open site, though, refers to all of the variables associated with a physically open area. Within these definitions, a population can be considered space limited only if an increase in the availability of space—but with no changes in any other variables—leads to a greater rate of population increase.

For communities of sessile aquatic organisms such as intertidal and stream invertebrates, to test for space limitation may be as simple as removing some individuals from the substrate. If the population is space limited, its per capita rate of change should increase. However, for plant communities, such an imposed disturbance would not test for space limitation. The removal of plants does increase the amount of the resource, space, but it also increases the availability of sunlight, soil nutrients, and water. If the per capita growth rate of a plant population were to increase after the removal of some individuals, this would imply that the population was limited by open sites, but not necessarily by space. In fact, it seems unlikely that space is a limiting resource in most plant communities. Most have sufficient physically open space that more individuals could occupy an area if other resources were in greater abundance. This, however, is not so for many marine invertebrate communities, such as intertidal and coral communities, in which organisms are densely packed and in which physical space for colonization and individual growth may be a limiting resource.

It would be possible to consider open sites as a limiting resource in plant communities, and to make some qualitative predictions about the relationship between the rate of supply of open sites and the diversity of communities. This, essentially, is what was done by Grubb (1977) in his discussion of the importance of the "regeneration niche" in plant communities. In an analysis of competition for open sites, various individual open sites would be classified with respect to qualities such as size, time of disturbance, and orientation. I, however, am leery of a "high level" summary variable such as "open sites", for I believe that it may obscure many important processes determining the structure of plant communities. In most plant communities, the important qualities of an open site are probably the light and nutrient availabilities associated with the site, and not the physically open space. I believe it will be more productive to consider light and each nutrient explicitly as resources than to lump these together in a summary variable. Although the dichotomy between closed sites and open sites may be useful in some situations, it ignores the complete gradation of conditions that occurs as microsite succession takes place. Associated with microsite succession are changes in the availability of soil nutrients and in the light available for seedlings and shoots. At least within the framework of a mechanistic approach to community structure, it is best to consider explicitly those factors for which competition is directly occurring. For these reasons, I will not discuss competition for open sites, but will instead consider competition for space as a resource. As such, the following, admittedly speculative discussion applies to communities of sessile marine animals, for which space may often be a limiting resource. It may apply to interactions in some plant communities, if there are plant communities which are truly space limited. However, it seems more likely that open-site limitation in terrestrial plant communities reflects light and nutrient limitation. This does not imply that disturbance is an unimportant process in terrestrial plant communities. Disturbances create spatial and temporal variance in the availability of limiting resources, especially light. The average disturbance rate that a plant community experiences should be closely related to the average rate of supply of light to the soil surface, and thus to seedlings and plants of small stature. As discussed in Chapter Five, the average rate of resource supply and the spatial variance in supply are important determinants of plant community diversity.

I will make several assumptions in this discussion of space as a resource. First, I assume that the disturbance processes which create new open space are such that openings will have a certain average size and that the variance in the size of openings is constant, independent of the rate of disturbance. If this is the case, the average amount of open space can be used as a measure of the availability of this resource. A simple variable, such as the percent open space, can summarize resource availability. Second, I assume that there are no seasonal trends in either disturbance patterns or in requirements for space. Clearly, a more complex treatment of space as a resource is possible. However, it is often instructive to start analyses with the simplest possible approaches to determine to what extent major patterns may be explained by general hypotheses. After the discussion of space as a resource, I will consider the general effects of disturbance on communities, including plant communities, for which space, by itself, may not be limiting, but for which disturbance may influence the availability and heterogeneity of limiting resources such as light and nutrients.

Space as a Resource

To consider open space as a resource, it is first necessary to explore the dependence of the long-term per capita growth rate of a species on the amount of physically open space. If space is a resource, the reproductive rate of the population must increase as the amount of unoccupied space increases, through some range of availability. Several different hypothetical response patterns are illustrated in Figure 79A. Some species may be unable to reproduce in areas without at least a certain threshold amount of space, as shown for species *D*. Other species may require little space for growth, and may be inhibited by lack of cover, such as species *A* of Figure 79A. Still other species may be intermediate in their requirements for space, such as species *B* and *C*. If the disturbance and colonization processes in a habitat were such that the average amount of available space were at the point numbered 1 on the *x*-axis of Figure 79A, species *A* would be dominant in the community, for it would have the highest growth rate. Similarly, a community with these four species which experienced a higher disturbance rate, such that the average amount of available space was at point 2, would be dominated by species *B*. Even higher disturbance rates could lead to average amounts of available space at points 3 and 4, which would be dominated by species *C* and *D*, respectively. These equilibrium points occur because increased disturbance rates necessarily imply increased mortality rates.

Secondly, for space to be a resource, space must be consumed. The acts of colonization and growth on a site constitute the consumption of space. The consumption rate of space by a species depends on its ability to colonize available space and on its rate of growth after colonization, and could be measured by the rate of disappearance of available space caused by a given species.

May (1973) discussed the problem of modeling population dynamics assuming population density could be represented by a non-integer variable, and showed that this could be an

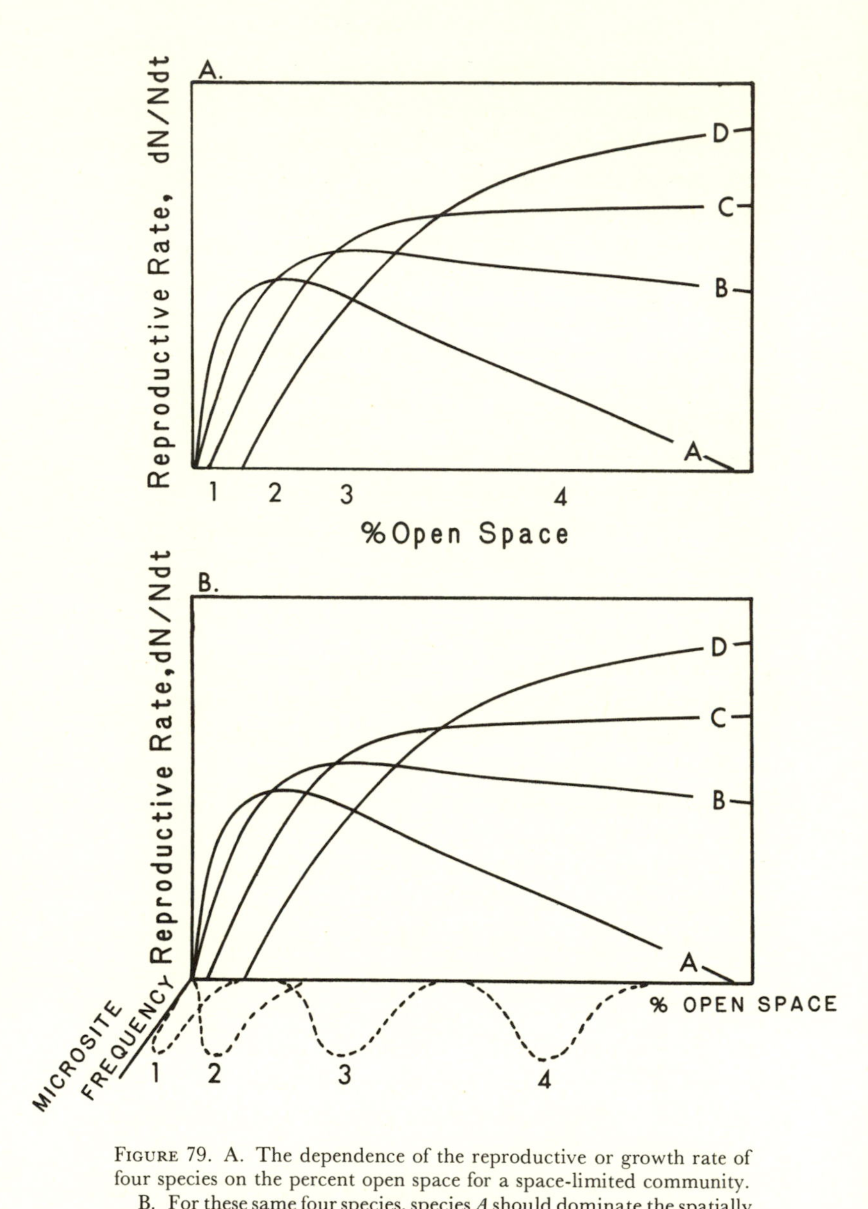

Figure 79. A. The dependence of the reproductive or growth rate of four species on the percent open space for a space-limited community.
B. For these same four species, species *A* should dominate the spatially variable habitat 1, species *A* and *B* should coexist in habitat 2, etc.

acceptable approximation for large population densities. However, for small population densities (less than about 100 individuals), the necessary restriction of population size to integer (whole organism) values led to considerable variance in the estimated population density compared to the less realistic model which assumed non-integer population sizes. May (1973) termed this variance "demographic stochasticity." A similar problem occurs when space is considered as a resource. A new area of available space occurs as the result of the removal of one or more individuals, and such an opening is eventually taken over by one or more new individuals. Even with a perfectly fixed average rate of disturbance, the discrete nature of individuals means that disturbance must always lead to variance around the average amount of available space in a habitat, for such sites are created by the loss of one, two, three or some other whole-integer number of individuals. This "spatial stochasticity" means that it is unrealistic to represent the amount of open space in a habitat as a fixed percent without also including the amount of microsite to microsite variability experienced by propagules of a given species. The microsite of a space-limited organism is defined as the area colonized by the majority of the offspring of an individual. Such variability can be represented as in Figure 79B in which an additional axis is added to show the probability distribution of the percent open space for a given site. This more realistic representation shows that habitat 1 of Figure 79A may still be dominated by just species *A*, while habitat 2 may have sufficient microsite to microsite variability in the percent open space to allow species *A* and *B* to coexist. Similarly, habitat 3 may have sufficient spatial stochasticity to allow species *B* and *C* to coexist, while habitat 4 may still be dominated by species *D*. This simple analysis of space as a resource may seem to suggest that, even for situations in which space is the only limiting resource, different rates of disturbance could lead to a humped diversity curve. Assuming individual growth curves such as in Figure 79, moderately low rates of disturbance would lead to coexistence of the greatest number of species, whereas habitats with very low disturbance rates (such as habitat 1) and habitats with very high disturbance rates (such as habitat 4) would be dominated by a few species. Such a conclusion, however, is of questionable validity because it depends on the exact placement of the growth curves in Figure 79. It is possible to choose growth curves for which this pattern does not occur. However, Figure 79B does illustrate a major difference between space and other resources. The discrete nature of the disturbance (and thus, mortality) processes necessarily imposes spatial heterogeneity in the availability of space, and such heterogeneity is capable of allowing several species to coexist, at equilibrium, on a single limiting resource.

Space and Food

Those species for which growth rate increases with the amount of open space may be space limited in their natural environment. However, other resources may also be in low availability, and thus may also limit growth. Because of this, it is necessary to know how space and other limiting resources such as food or nutrients jointly affect the growth rate of a species. Such resources are most likely essential, giving growth isoclines with one of the two shapes of Figure 2E and F. That they are essential implies that no amount of one resource can completely substitute for a lack of the other. It seems likely, though, that space will be an interactive essential resource (Fig. 2F) relative to food or nutrients. This means that one resource can partially substitute for the other resource at intermediate availabilities of the two resources. For the rest of this analysis, I will assume that space and food are interactive essential resources, i.e., I will use ZNGI's with curved corners. This assumption, however, has no effect on the community level prediction to be made later in this chapter. By explicitly considering both open space and another resource, this analysis allows for openings which differ in resource availability.

Competition

Because of the spatial and temporal scales on which the process is usually observed, competition for open space may seem to be a more complex process than competition for other resources. One prevailing view of the effect of disturbance on community structure is that disturbance allows inferior competitors to coexist with superior competitors if the inferior competitors are superior in their ability to colonize newly disturbed sites. This is a view of communities as a mosaic of patch types, with competition in any one patch type leading to a successional transition from species which are superior colonizers to species which are superior competitors. Although these are the underlying processes assumed in the equilibrium model of competition for space and a food resource developed in this section, the model explicitly includes competition for space and competition for other resources, and is thus counter to the view that disturbance decreases competitive interactions. The model presented here establishes the conditions under which such processes can lead to the long-term, equilibrium coexistence of species.

The ZNGI's of Figure 80A illustrate an equilibrium model of competition for space and a food resource. As drawn, species *A* has a higher requirement for food but a lower requirement for open space than does species *B*. Thus, species *B* would tend to dominate a site just after a disturbance, and species *A* would tend to dominate the site once it was more closed. Species *B* is the superior competitor for space, able to establish and maintain a stable population in a relatively closed area, whereas species *A* is the superior competitor for food. A habitat with the resource supply point labeled 1 would be relatively food-rich but have a low disturbance rate. Such a habitat would

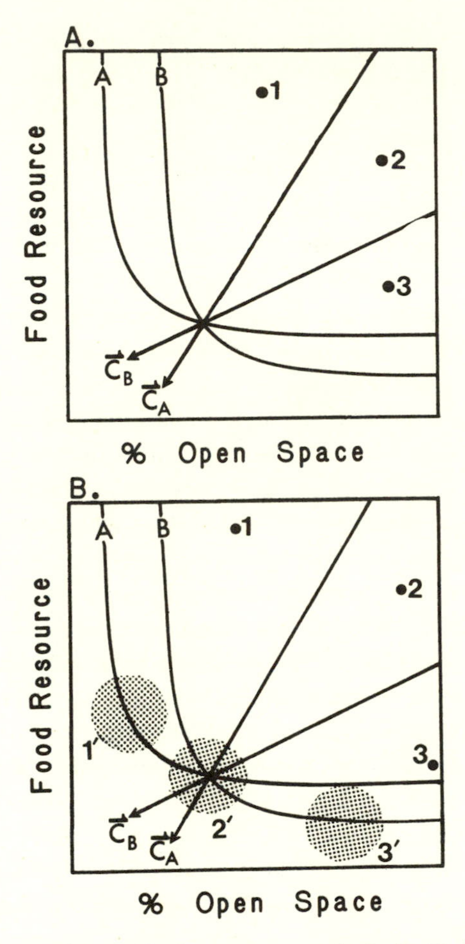

FIGURE 80. A. Competition between two species for limiting space and a limiting food resource. This case may apply to situations of competition among sessile marine animals. The requirements of these two species for these resources define regions in which species *A* is dominant, both species coexist, or species *B* is dominant.

B. Because of spatial stochasticity, a given rate of supply of open space (i.e., a given disturbance rate) should lead to variance even in a habitat in which these processes occur at a uniform rate. This occurs because of the discrete nature of individuals in the population.

be dominated by species *A*, with species *B* tending toward extinction because species *A* would reduce the availabilities of the two resources to a point inside the ZNGI of species *B*. Habitat 3, with a high disturbance rate and a low rate of supply of food, would be dominated by species *B*. An intermediate rate of supply of food and an intermediate disturbance rate (habitat 2) would lead to coexistence. At the equilibrium point, species *A* would be more limited by the food resource and species *B* would be more limited by the amount of open space, giving a stable equilibrium. The long-term effect of consumption by these species in habitat 2 would be that the availability of food and open space, *averaged over numerous microhabitats*, would be at the point at which the two ZNGI's cross, the two-species equilibrium point. However, the spatially discrete nature of disturbances would mean that microhabitats would differ in their amounts of food and open space even when the whole system was at equilibrium. Thus, supply point 2 might lead, at equilibrium, to microhabitats which have the distribution shown around the two-species equilibrium point in Figure 80B, while supply points 1 and 3 would lead to the other distributions shown. Figure 80B thus illustrates the difference between a micro-view of these processes and their long-term effect on community composition.

Figure 81A shows a case of four species competing for these two resources. If the spatial stochasticity imposed by disturbance were small, a habitat experiencing a fixed disturbance rate and a fixed supply rate of food (point 1) would be dominated by a pair of species, with the other species going extinct. If, however, disturbance-caused heterogeneity (i.e., spatial stochasticity) were much greater, the same supply point could allow the coexistence of all four species, as shown in Figure 81B.

It is possible for space as a resource to allow the long-term coexistence of more species than there are resources even when a habitat is homogeneous in the processes that supply space, because these processes necessarily occur in discrete places at

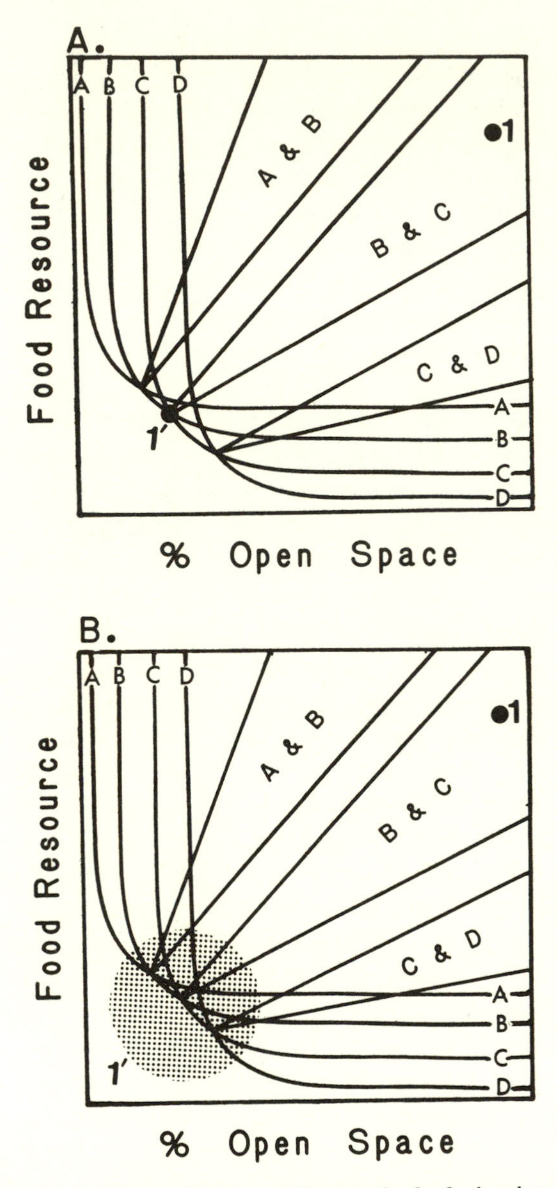

FIGURE 81. A. Competition among four species for food and space. In the absence of spatial stochasticity, only species *B* and *C* could coexist in habitat 1.

B. With spatial stochasticity, all four species would be able to coexist in habitat 1. Species *B* and *C* would still be the dominant species in the habitat.

discrete points in time, and thus create heterogeneity. The supply process for foods, nutrients, and other resources is not so discrete. Although they are all consumed as single entities by an organism, a single individual often consumes many such entities, and thus its reproductive rate represents the integrated effect of what can often be considered continuous variables. However, for space, the reproduction of an individual depends on whether or not it reaches an opening, not on the average number of openings in a much larger area.

Disturbance and Species Diversity

The isocline approach of Figures 80 and 81 may be used to ascertain the possible effects of various rates of supply of space and food on the diversity of plant communities. Such an analysis is similar to that presented in Chapter 5, except for one complication. In the previous analysis it was assumed that changes in the supply rates of nutrients did not affect the mortality rate experienced by the species. Although most nutrients in plant communities are supplied via recycling, it seems a valid approximation to assume that resource supply is not directly coupled with mortality rate. However, a new opening can be supplied only when some individual dies. Thus, changes in the disturbance rate must be accompanied by changes in the mortality rate of various species. Such changes in the mortality rate mean that ZNGI's will shift as disturbance rate changes, since a species requires more resources to maintain a stable population when it experiences higher mortality rates.

This is illustrated for three possible cases in Figure 82. In all cases, the average disturbance rate increases from 0.1 to 0.2. However, in some cases, all species are equally affected by disturbance, whereas in other cases some species are more susceptible to disturbance-caused-mortality and other species are less susceptible. For Case 1, an increase in disturbance rate affects all species equally. All the ZNGI's are shifted equally far from the origin at the higher disturbance rate. This is

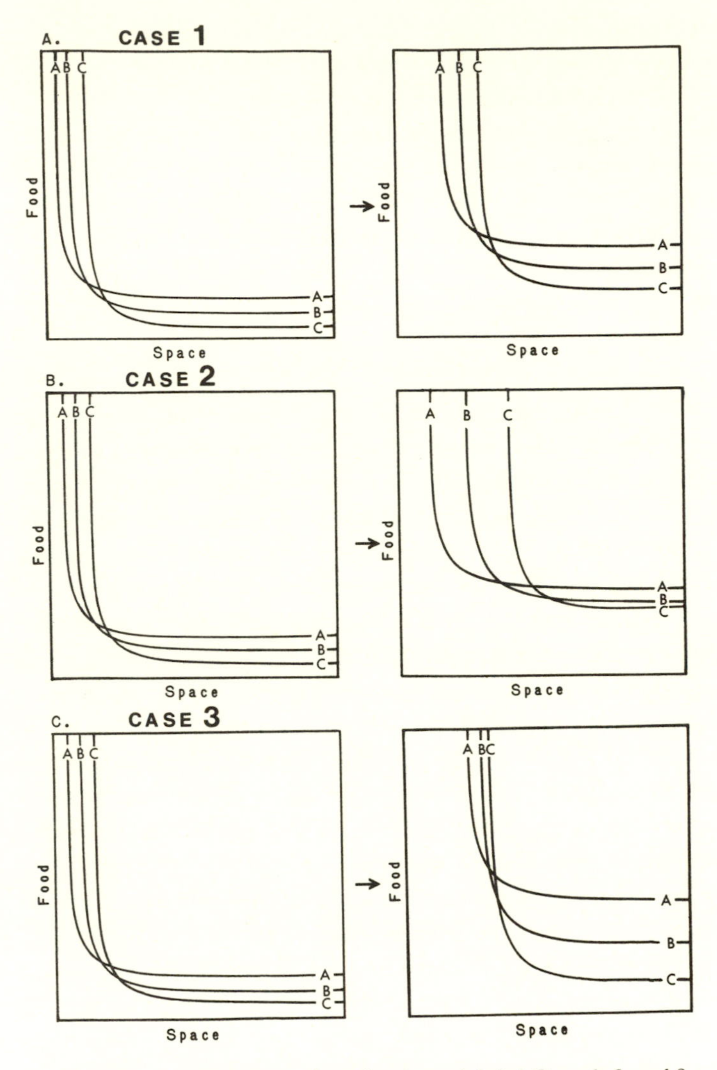

FIGURE 82. The three sets of graphs above, labeled Cases 1, 2, and 3, illustrate the three qualitatively different ways that an increase in disturbance rate may affect the placement of the ZNGI's of three species. For Case 1, the increased mortality caused by disturbance falls equally heavily on all species, causing all their ZNGI's to be shifted away from the origin. For Case 2, mortality falls most heavily on species *C* and least heavily on species *A*, leading to the pattern shown. For Case 3, species *A* experiences the most disturbance-caused mortality.

comparable to having non-selective disturbance, herbivory, or predation. For Case 2, the increased disturbance causes changes in mortality which fall most heavily on species C, less on species B, and least on species A. Such a pattern of disturbance-caused-mortality is one in which the species which has the highest requirement for open space (i.e., the species which tends to dominate disturbed sites) is also most susceptible to disturbance-caused-mortality. For Case 3, the species which is the best competitor for limiting space suffers the most mortality from disturbance.

I explicitly modeled several cases of competition among twenty species for limiting space and nutrient resources. For all cases, I assumed that the parameters describing the resource requirements of each species were unchanging. The positions of the ZNGI's of the twenty species for a case in which all were experiencing the same mortality rate are illustrated in Figure 83. As shown in this figure, species 1 is the superior competitor for open space and the poorest competitor for food. Species 20 is the poorest competitor for open space and the best competitor for food. The other eighteen species are intermediate in their requirements for the two resources. All the species were assumed to have the same maximal growth rate. As before, a bivariate normal distribution was used to describe the microsite to microsite spatial variation in the supply rate of the nutrient and the disturbance rate. Spatial stochasticity was assumed to be included in the microsite to microsite variation of the bivariate normal distribution. If it had been included separately, a few more species would have been able to coexist at any given disturbance rate, but the overall relationship would not have been changed.

Three different cases were explored. In Case 1, all species were equally affected by the disturbance rate. Thus, an increase in disturbance rate for Case 1 would shift the ZNGI's of all species equally away from the origin, as shown in Figure 82A. In Case 2, the species which was the best competitor for

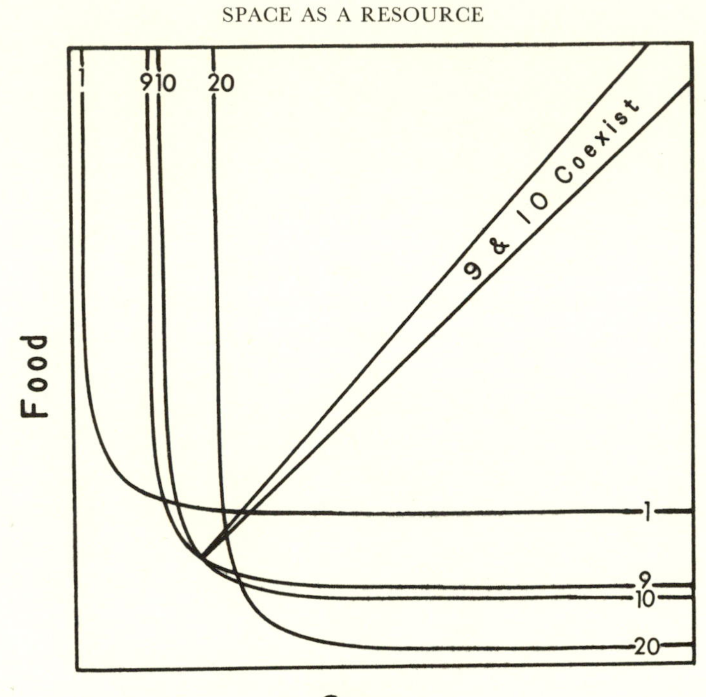

FIGURE 83. This shows the qualitative placement of the ZNGI's of the 20 species used in the simulation of competition for space and food in a spatially structured habitat. Species 1 is the superior competitor for space and species 20 is the superior competitor for food.

open space (species 1) was assumed to suffer the least disturbance-caused mortality, with the species that was the best competitor for food (species 20) suffering most of the mortality caused by increases in the disturbance rate. The other species suffered intermediate increases in mortality rate with increases in disturbance rate. For Case 3, the best competitor for open space (species 1) suffered most of the mortality from disturbance, whereas the species which was the worst competitor for open space (species 20) suffered the least mortality from disturbance. In all cases, a background mortality rate of 0.33

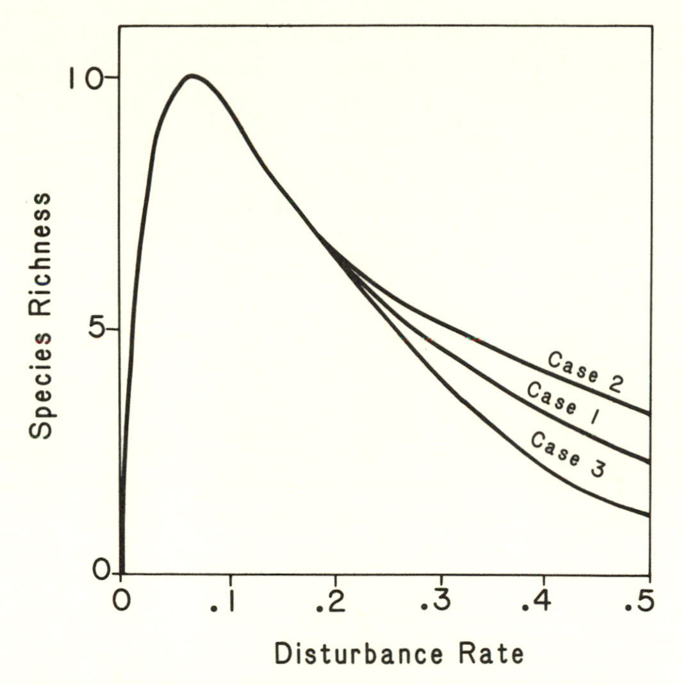

FIGURE 84. The predicted dependence of species richness on disturbance rate for three cases similar to those in Figure 82. For cases in which disturbance-caused mortality falls equally on all species, independent of their competitive ability for space (Case 1), and for cases in which the superior competitor for space experiences the least disturbance-caused mortality (Case 2), and for cases in which the superior competitor for space experiences the most disturbance-caused mortality, species richness is predicted to give a humped curve when graphed against disturbance rate. The differences between these three curves are minor and occur only at low species richnesses.

of the maximal reproductive rate was assumed, and all species were assumed to have the same maximal reproductive rate. Half of the increase in mortality caused by increased disturbance was assumed to substitute for the background mortality rate.

Figure 84 shows how changes in the disturbance rate for Cases 1, 2, and 3 influence the equilibrium species richness of these communities. The disturbance rate shown on the x-axis of

Figure 84 is an average over all the species for a given spatially heterogeneous habitat. For all three cases, the species richness–disturbance rate curve is humped, with maximal diversity in habitats with moderate disturbance rates. Compared to the case in which all species experience the same mortality rate (Case 1), species diversity tends to fall less quickly with disturbance rate when the species with the highest requirement for open space suffers the most mortality (Case 2), and tends to fall more rapidly when the species with the highest requirement for open space suffers the least mortality (Case 3). These differences are slight, however, and the curves never differ by more than a few species. In all three cases, changes in the supply rate of the resource space—i.e., changes in the disturbance rate—affect species richness in the same way as do changes in the supply rate of other essential resources (compare Figs. 38 and 84).

Disturbance and Diversity

These results apply only to communities in which open space is a limiting resource, and thus probably do not apply to terrestrial plant communities. As already discussed, it seems likely that light, but not space, is a limiting resource in terrestrial plant communities, especially those with relatively high biomass or a closed canopy. However, when these results are combined with those presented in Chapter 5, they suggest that disturbance should have a similar effect on the diversity of plant communities as it is predicted to have on the diversity of communities of sessile marine invertebrates. In the case of sessile marine organisms, disturbance may be directly supplying a limiting resource, space. In the case of plant communities, disturbance indirectly supplies a limiting resource, light, by removing individuals that had been capturing the light because of their stature. In addition, by removing these individuals, disturbance removes potential competitors for soil nutrients and water, and thus leads to increased rates of availability of these potentially limiting resources. If disturbance in terrestrial plant communities is thought of as a process that supplies essential resources—light and nutrients—the theory in Chapter 5 predicts that the species richness of this type of community should depend on disturbance rate just as it should depend on any other measure of resource richness. Thus, we would expect that the diversity of a plant community would be maximal at moderately low rates of disturbance, and decrease for either increases or decreases in the disturbance rate.

Many others have suggested that periodic disturbance, including predation, may be important in maintaining the species diversity of prairies (Platt and Weis, 1977), tropical forests and coral reefs (Connell, 1978; Huston, 1979), intertidal sessile animal and plant communities (Paine, 1966; Lubchenco, 1978), and plant communities in general (Grubb, 1977). An excellent experimental study by Lubchenco (1978) demonstrated that the species richness of macrophytic algae in marine tidal pools was maximal at intermediate rates of herbivory, and lower for either greater or lesser rates of herbivory. Her data are shown in Figure 85A. This pattern is consistent with that predicted by the theory presented, if it is assumed that disturbance is providing light and possibly decreasing the magnitude of nutrient competition. Lubchenco also studied the effects of various levels of herbivory on the macrophytic algal communities on emergent substrata (rocks), with her results shown in Figure 85B. Lubchenco (1978) suggested that the data points in Figure 85B were best fit by a straight line, which would indicate that species richness was maximal on the emergent substrata in the absence of herbivory. She hypothesized that the difference between the tidal pools and the emergent substrata (i.e., the difference between a humped diversity curve and a linearly decreasing diversity curve) was caused by differences in the competitive ability of the algae in the two habitats, with no comparable change in the dietary preferences of the herbivorous snails. In tidal pools, the herbivorous snails

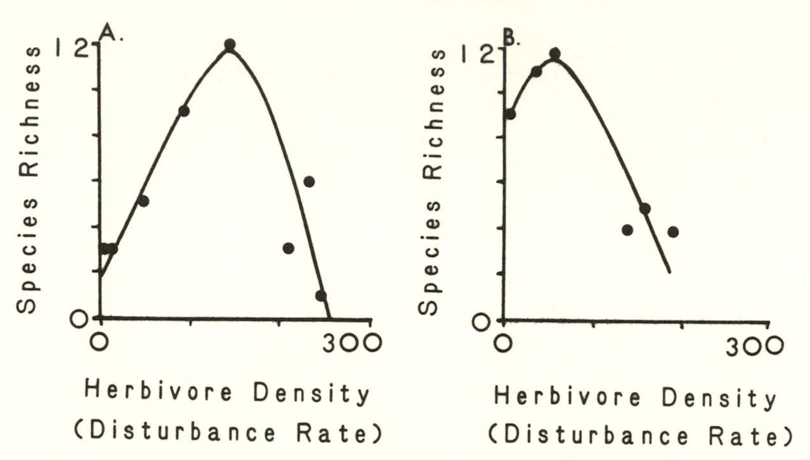

FIGURE 85. The observed dependence of plant species richness on herbivore density for marine intertidal pools (part A) and for emergent substrata (part B), as observed by Lubchenco (1978). The curves drawn through the data show my interpretation of Lubchenco's results.

preferentially consumed *Entermorpha*, the plant which Lubchenco said was the competitive dominant for space (light). On the emergent rocks, the snail still preferred *Entermorpha*, but it was no longer the competitive dominant. This led Lubchenco to suggest that disturbance (herbivory) which selectively removed the superior competitors for the limiting resource would lead to a humped diversity curve, whereas disturbance which selectively removed inferior competitors for the limiting resource would lead to a linear relationship. This intuitively appealing hypothesis is not consistent with the theoretical predictions shown in Figures 38 and 84. For nonselective disturbance, as well as for disturbance which selectively removes either the competitive dominants or the competitive subordinants for space, the theory developed in this chapter predicts a humped diversity curve. The interesting hypothesis suggested by Lubchenco (1978) needs further theoretical exploration to determine whether there are conditions for which a linear relationship would be obtained. I might suggest an alternative hypothesis for the data of Figure 85B. The plants on the emergent rocks may have been experiencing a greater background rate of non-herbivore disturbance, such as from wave action, than the plants in the more protected tidal pools. The background disturbance rate on the emergent substrata may have been such that, in the absence of herbivory, the community was just below the diversity peak. If this were so, increased rates of snail-caused disturbance would be expected to give a humped curve which was truncated at a "veil line." The curve I have drawn through Lubchenco's data (Fig. 85B) illustrates this alternative view.

The usual explanation for the effect of disturbance on species richness is that disturbance prevents competitive displacement by opening up areas which are colonized by inferior competitors capable of reproduction before displacement can occur. This is a view that has disturbance ameliorating competition, allowing a mosaic pattern of hopscotch coexistence. Connell (1978) and Lubchenco (1978) have suggested that disturbance, acting in this fashion, should lead to communities with a humped diversity curve when species richness is plotted against disturbance rate. The results presented in this chapter suggest that a graph of species richness against disturbance rate should give a humped curve, but for somewhat different reasons than suggested by Connell, Lubchenco, and others.

According to the ideas discussed in this chapter, only the very best competitors for open space or light can survive in habitats with very low disturbance rates. These are the species which can become established in almost closed communities. Increases in disturbance rate allow such a community to be invaded by other species which are superior competitors for another limiting resource, but are inferior competitors for open space or light. Thus, communities with moderate rates of disturbance can have numerous species coexisting. However, further increases in disturbance rate preclude establishment of those species which are the best competitors for open space or light because the species which are the best competitors for another limiting resource reduce the level of that resource below the point at which the superior competitors for open space or light can survive. This view of the role of disturbance in determining the structure of communities of sessile organisms differs from that of Connell (1978) and Lubchenco (1978) in that it does not consider disturbance to be a process that periodically interrupts competition, but rather it considers disturbance to be a process that influences the relative supply rates of the resources for which competition occurs. For terrestrial plant communities in which disturbance indirectly supplies light and possibly nutrients, this view of disturbance suggests that its effect on diversity comes from changes in resource richness and resource ratios in a habitat, and not from any amelioration of competitive effects. The species composition of a light and nitrogen limited plant community should be strongly influenced by the ratio of the disturbance rate to the supply rate of nitrogen (assuming that disturbance mainly supplies light), just as the species composition of Figure 49 is determined by S_1/S_2, the ratio of the supply rates of its limiting resources.

SUMMARY

The view of space as a resource developed in this chapter suggests that the major qualitative effects of differences in the supply rate of space (i.e., disturbance) are the same as the major effects of differences in the supply rate of any other limiting essential resource. This is so despite the complication that an increased supply rate of space necessarily means an increased mortality rate for at least some populations in a community. This result adds credence to the predictions of Chapter 5, for it suggests that relaxation of the assumption that resource supply is independent of mortality will not change the qualitative trends predicted. Even if increased rates of nutrient supply for a plant community could only occur through increased death rates of the occupants and recycling of their nutrients, the analysis of this chapter suggests that a humped diversity curve would still be expected. Similarly, the results presented here and in Chapter 5 suggest that disturbance—whether it is supplying the limiting resource, space, or indirectly allowing increased supply rates for other essential resources such as light or various nutrients—should lead to maximal diversity in communities with moderately low disturbance rates.

For space-limited communities such as marine intertidal communities, disturbance rate may be a good measure of the supply rate of a limiting resource, space. As such, there should be potentially predictable changes in the species composition of these communities in response to differences in disturbance rate. In contrast, for terrestrial plant communities, apparently similar disturbance events may differ greatly in the amounts of light and nutrients that are provided, and the open space provided by disturbance may be irrelevant. Although the diversity of terrestrial plant communities should depend on disturbance rate because disturbance rate is a measure of resource richness, it would seem better to study explicitly the resources supplied by disturbance. It is only from information on the actual rates of supply of these various resources that the mechanisms leading to the dominance and diversity patterns in terrestrial plant communities may be ascertained.

CHAPTER NINE

Concluding Questions and Speculations

The main purpose of this book has been to build an equilibrium theory of competition for limiting resources and to explore some of the implications of such a theory for the species composition and diversity of communities. There are many important questions about resources and community structure not covered in this book. In this chapter I will briefly explore a few additional questions which intrigue me, and offer some speculative answers.

PLANT SUCCESSION

Ecologists have long recognized regular, repeatable, and apparently directional patterns of change in the species composition of plant communities. Such changes have been termed species succession. From the pioneering work of Cowles (1899) and Cooper (1913) arose one of the first major ecological theories, the theory of succession proposed by Clements (1916). The simplicity, completeness, and purpose contained in Clement's theory were sufficiently satisfying to most ecologists that it survived numerous attacks (e.g., Gleason, 1917, 1927; Egler, 1954) and remained the most commonly held view of succession until the question was reexplored in papers by McCormick (1968), Odum (1969), Drury and Nisbet (1973), Colinvaux (1973), Horn (1974), Connell and Slatyer (1977), and others. For brevity, the hypotheses proposed by these researchers will not be reviewed here, but I would like to propose an alternative view of succession, which may be called the "resource ratio" hypothesis.

Although the particular details of plant community succession are unique to each site, there are some general trends that are common to many communities. Such trends may be most easily understood by considering the successional sequence that follows a major disturbance to an area, such as fire, cultivation, a landslide, or the retreat of a glacier. The repeatable changes in species composition following such a disturbance are generally associated with increased nutrient richness of the soil and increased competition for light. Consider, for instance, the primary succession following glacial recession in Glacier Bay, Alaska (e.g., Cooper, 1939; Crocker and Major, 1955; Lawrence, 1958). The till left by the receding glaciers is a very nitrogen-poor mineral soil. Most of the plants initially colonizing this soil are stunted, having the yellow, chlorotic leaves indicative of severe nitrogen deficiency. Only a few species are capable of successful reproduction on this very nitrogen-poor soil. Many tree species, such as willows and cottonwoods, have a stunted, shrublike morphology. Soil nitrogen levels increase immensely during the first 100 years of succession, mainly because of the nitrogen fixation capabilities of the alders and *Dryas* which form dense stands on the low nitrogen soil (Crocker and Major, 1955; Lawrence, 1958). As soil nitrogen levels increase, willows and cottonwoods become erect trees and overshade the alders. This stand is then invaded by spruce which eventually form a closed canopy above the alder. The spruce stand is then invaded by two species of hemlock which coexist with the spruce. Thus, during succession, there is a change from a habitat which is low in nitrogen but has a high availability of light for seedlings and shoots to a habitat which is high in nitrogen but has a closed canopy.

The basic process of primary succession at Glacier Bay can be represented by Figure 86, in which soil nitrogen and the amount of light available at ground level are assumed to be the two

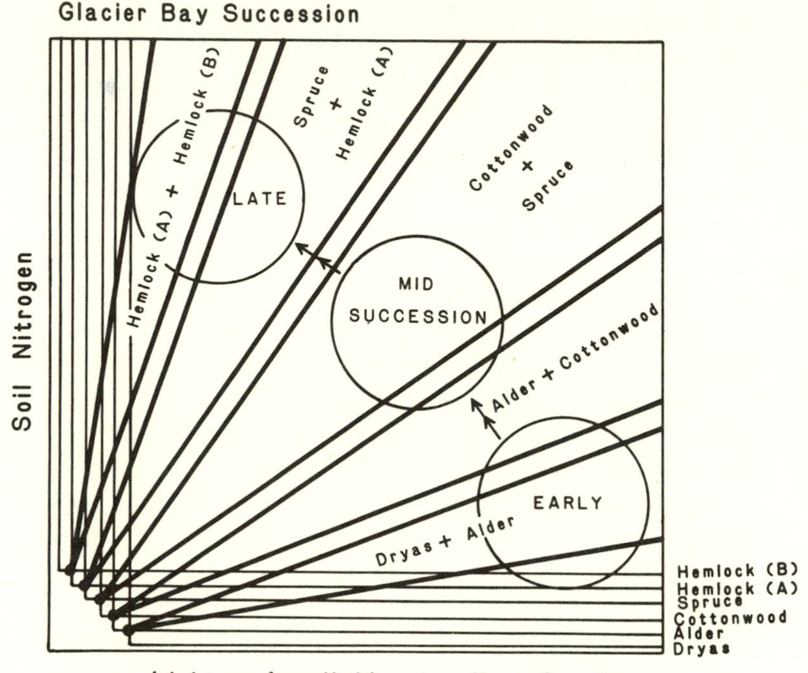

FIGURE 86. The hypothesized positions of the ZNGI's of some of the major species (or groups of species) in the successional sequence observed at Glacier Bay, Alaska. This figure illustrates the hypothesis that succession occurs as the relative availabilities of light and nitrogen change in response to soil development. Nitrogen-fixing plants are shown as having the lowest requirement for nitrogen but the highest requirement for light, and the two species of hemlock are shown as having the highest requirement for nitrogen but the lowest requirements for light (at the ground level).

limiting resources. Hypothetical growth isoclines are shown for six of the major taxa, as well as changes in the relative availability of the two resources through time. As an area goes from being a low nitrogen, high light habitat to a high nitrogen, low light habitat, a successional sequence occurs because of the differing requirements of these species for these resources. Figure 86 illustrates the "resource ratio hypothesis," the hypothesis that the successional sequence is determined by changes in the ratios of the limiting resources.

A similar explanation may be proposed for many other successions. For any plant community in which there are at least two limiting resources, succession is expected if the relative supply rates (i.e., the ratios) of the resources change through time. For many freshwater lakes, phosphate, silicate, and nitrate are the most limiting resources. Consider species with phosphate and silicate requirements similar to those of the diatoms of Figure 87. Early in the spring, after spring turnover, phosphate and silicate are high in abundance. The consumption of these nutrients by growing algal populations decreases their availability. Because phosphate is readily resupplied by within-water-column processes and silicate is not, the relative availabilities of these two resources will change. The supply rate of silicate drops much more quickly than that of phosphate. Thus, there should be a spring successional sequence among diatoms from those which are superior competitors for phosphate to those which are superior silicate competitors. The relationships in Figure 87 lead to a successional sequence which is consistent with patterns reported by Kilham (1971). As diatoms become increasingly limited by silicate, species of green algae which are superior phosphate competitors compared to the remaining diatoms should become dominant. If nitrogen becomes limiting, the green algae and blue-green algae which are superior competitors for nitrogen should tend to dominate. Additionally, as total algal biomass increases, light becomes increasingly limiting. This suggests that the complete seasonal successional sequence in a lake may be determined by changes in the proportions of three or four limiting resources, and may be a complex puzzle to piece together. However, knowledge of the resource requirements of the major species and of the changing patterns of the supply rates of limiting resources may allow prediction of the successional sequence.

This discussion of succession points out the necessity of a synthesis of the often disparate approaches of population and

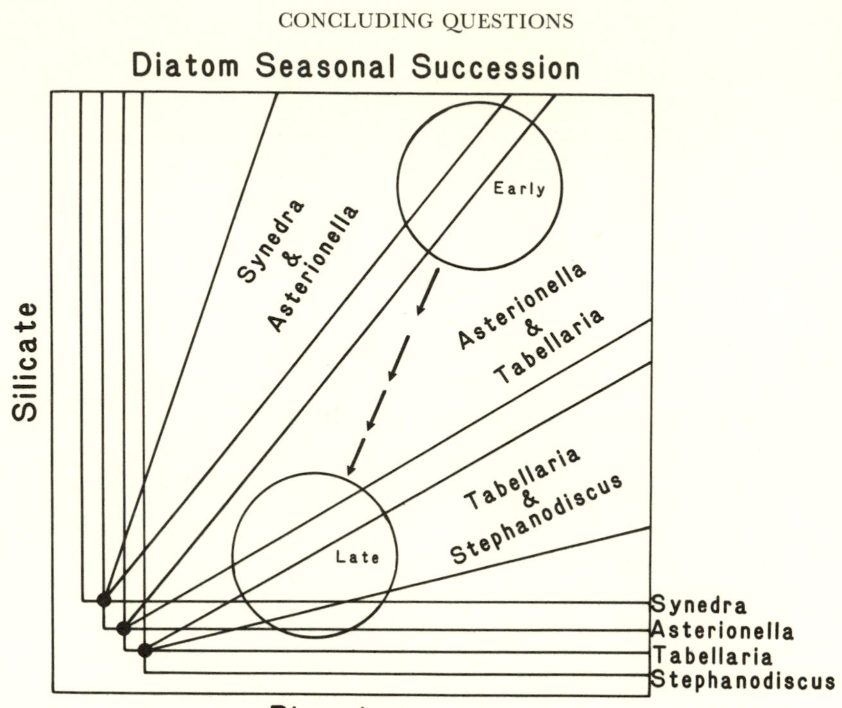

FIGURE 87. The relative positions of the ZNGI's of four species of freshwater diatoms are shown for a situation in which phosphate and silicate are the two limiting resources. These isoclines predict a successional sequence from *Synedra* and *Asterionella* to *Tabellaria* to *Stephanodiscus* as silicate becomes increasingly limiting. This prediction is qualitatively consistent with Kilham's (1971) observations on silicate and the diatom successional sequence.

ecosystem ecology. If some of the trends in a successional sequence are caused by changes in the relative rates of supply of limiting resources, it will only be possible to understand succession when information on the resource requirements of the competing species is combined with knowledge of the processes determining the budgets and rates of supply of the limiting resources. This will require that intensive studies of single species be combined with intensive studies of whole-system processes. Without knowledge of the processes deter-

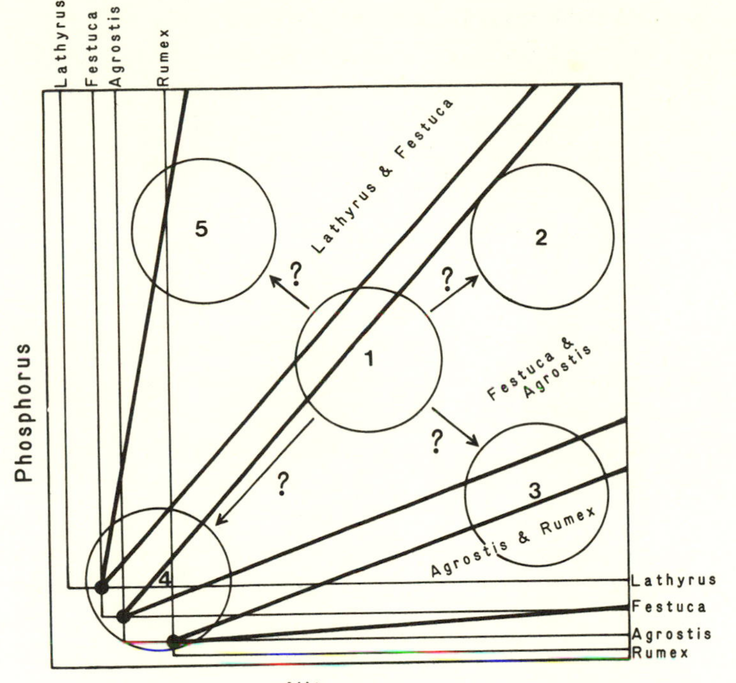

FIGURE 88. The ZNGI's of Figure 68 are used to illustrate the need for a synthesis of the mechanistic approach to resource competition with the more traditional ecosystem approaches to nutrient budgets and nutrient cycling. If a habitat were at point 1 at a given period in time, it would be impossible to predict the future species composition and diversity of this community through time unless it were known how these and other species in the community, as well as outside forces, were influencing the rates of resource supply. The successional pattern observed, if any, from habitat condition 1 into the future requires a marriage of the mechanistic approach to community structure with an ecosystem view of nutrient processes.

mining the budgets of various nutrients, it would be impossible to predict either the direction or the rate of succession. This is illustrated in Figure 88 for a hypothetical case of succession in a community of N and P limited plants. If this plant community had resource supply processes that put it at point 1 at a particular time, it would be impossible to make any predictions about the future state of the community without knowledge of the processes determining the budgets and rates of supply of the limiting nutrients. This brief theoretical treatment of succession suggests that a synthesis of population and ecosystem approaches may allow us to understand the universal process of succession.

HOW MANY SPECIES MAY COEXIST ON A RESOURCE?

Many authors have asserted that the number of species that can coexist in a spatially homogeneous, equilibrium community cannot exceed the number of limiting resources (Volterra, 1931; MacArthur and Levins, 1964; Levins, 1968; Armstrong and McGehee, 1980), or must be less than the number of "limiting factors" (Levin, 1970). The theory developed in this book has shown that there may be many more species than limiting resources in a spatially structured, equilibrium community. The relaxation of one assumption—the assumption of spatial homogeneity—has a major qualitative effect on the community-level predictions of a theory of consumer-resource interactions.

In his "Homage to Santa Rosalia" and again in his "Paradox of the Plankton," Hutchinson (1959, 1961) suggested that high community diversity might be maintained by a variety of factors, including spatial structure, trophic complexity, and non-equilibrium conditions. We have seen that the predictions of a simple model of consumer-resource interactions in a spatially homogeneous, equilibrium environment are dramatically changed if spatial structure is added. What would happen if we were to modify the simple consumer-resource model in the other ways suggested by Hutchinson?

A. Adding a Trophic Level

The consumer-resource model developed in this book is a model of a two-trophic-level community. How might the predicted species diversity of a spatially homogeneous, equilibrium community be affected if a third, predatory trophic level were added? The general form for differential equations governing a three-trophic-level model would be as follows:

RESOURCE DYNAMICS: (J equations)

$$dR_j/dt = g_j(R_j) - \sum_i c_{ij} N_i f_i(R_1, R_2, \ldots),$$

CONSUMER DYNAMICS: (I equations)

$$dN_i/dt = N_i f_i(R_1, R_2, \ldots) - N_i m_i - \sum_k s_{ki} P_k q_k(N_1, N_2, \ldots),$$

PREDATOR DYNAMICS: (K equations)

$$dP_k/dt = P_k q_k(N_1, N_2, \ldots) - P_k d_k,$$

where R_j is the availability of resource j, N_i is the population density of consumer species i, P_k is the density of predator k; g_j is the function describing supply of resource j, f_i is the functional dependence of the growth of species i on the availability of all resources, q_k is the functional dependence of the growth of species k on the population densities of their prey (the consumer species); c_{ij} is the per capita rate of consumption of resource j by species i, s_{ki} is the per capita rate of consumption of consumer species i by predator species k, m_i is the mortality rate of species i in the absence of predation, and d_k is the mortality rate of the predator species.

For a three-trophic-level system such as described above, with J resources, I consumer species, and K predators, there will be J, I, and K equations, respectively, for the rates of change of these variables. When solved for equilibrium ($dR_j/dt = dN_i/dt = dP_k/dt = 0$), these equations place limits on the number of species that can stably coexist at biologically meaningful equilibria. For instance, there are $J + K$ equations defining the unknown population densities of the consumer species. If the number of consumer species were greater than $J + K$, $J + K$ equations would have to be solved for more than $J + K$ unknowns. Although this is possible given particular values for these variables, the slightest change in any of the variables would lead to extinction of the number of consumer species in excess of $J + K$. Biologically meaningful equilibria will occur only when

$$I \leqslant J + K,$$

i.e., the number of consumer species must be less than or equal to the sum of the number of resources and the number of predator species. Levin, Stewart, and Chao (1977) have demonstrated that two bacterial strains can coexist on one resource in the presence of one predator (a virus). These experimental results are consistent with these theoretical predictions.

At equilibrium, there are K unknowns describing the densities of the predator species which are defined by I equations. Biologically meaningful equilibria can occur only when

$$K \leqslant I,$$

i.e., the number of predator species must be less than or equal to the number of consumer species.

On the surface, the inequalities for a three-trophic-level system may not seem very distinct from the inequalities for a two-trophic-level system. However, let us ask how they limit the diversity of a community which has a *single* limiting resource. With a single limiting resource (i.e., $J = 1$), one consumer species can invade ($I = 1$). The presence of the consumer will allow a predator to invade ($K = 1$). Now that a predator is present, a second consumer can invade because $K + I = 2$. The invasion of a second consumer ($I = 2$) would allow another predator to invade successfully ($K = 2$), which in turn would allow another consumer to invade, etc. The inequalities describing the number of consumer and predator species which can coexist in an equilibrium, spatially homogeneous habitat place *no limit* on the eventual species richness of an environment with just one limiting resource. An additional trophic level has a potentially major impact on the diversity of an equilibrium community, increasing the possible diversity from being no greater than the number of resources to being unlimited. Just as adding spatial structure to a simple two-trophic-level model can allow an unlimited number of species to coexist, so adding a third trophic level to the simple model can similarly allow coexistence of an unlimited number of species.

What would happen if the simple two-trophic-level model of consumer-resource interactions in a spatially homogeneous, equilibrium environment were modified to allow for non-equilibrium conditions?

B. Non-equilibrium Solutions

Expanding on a suggestion of Levin (1970), Robert Armstrong and Richard McGehee have published a series of papers dealing with the long-term non-equilibrium persistence of consumers in a two-trophic-level system (see summary and extensions in Armstrong and McGehee, 1980). Levins (1979) considered the same question from a different perspective. Armstrong and McGehee (1980) and Levins (1979) have shown that many more consumer species can stably persist, but not at fixed population densities, than there are limiting resources, if the growth functions of the consumers are non-linear (e.g., if the function f_i of Eq. 2 is a saturating function instead of the more biologically unrealistic straight-line function assumed in some earlier models). This long-term persistence of more consumers than resources is the result of a non-equilibrium solution to the models. The same models, if solved for equilibrium (i.e., for $dR_j/dt = dN_i/dt = 0$), predict that the number of coexisting species cannot exceed the number of resources. Thus, modifying the simple two-trophic-level model to include non-equilibrium conditions potentially allows an unlimited number of species to coexist on a few limiting resources in a spatially homogeneous habitat.

Levins (1979) suggests that there is a rule specifying how many species can coexist under non-equilibrium conditions on a given number of resources. He asserts that each distinct non-linearity in the functions describing consumer growth and resource supply creates fluctuations in resource availabilities, and that these fluctuations may be responded to as if they were resources or "limiting factors." He suggests that the number of consumers that can persist in a non-equilibrium community is less than or equal to the sum of the number of resources and the number of distinct non-linearities.

Levins (1979) and Armstrong and McGehee (1980) discuss cases in which two species stably persist when limited by one resource. These cases illustrates how variance in the availability of a single resource may be considered to be a resource or a "limiting factor." Figure 89 illustrates Levins' example. Species 1 responds to the availability of the resource in a linear manner, and species 2 responds to the resource in a sigmoid manner. If these species were to experience the mortality rate shown (m), species 1 would have a lower equilibrium requirement for the resource than species 2 (i.e., $R_1^* < R_2^*$). Thus, species 1 would competitively displace species 2 under equilibrium conditions. Under non-equilibrium conditions, a different result could occur. Fluctuations in the availability of R around R_1^* will have *no effect* on the long-term average growth rate of species 1. This is because its linear growth function means that its capita rate of growth will be increased by fluctuations above R_1^* by exactly the same amount as its per capita rate of growth will be decreased by fluctuations below R_1^*. However, fluctuations in R will *increase* the per capita reproductive rate of species 2. The approximately exponential shape of the growth function of species 2 in the vicinity of R_2^* means that fluctuations in resource levels above the average will increase growth rate more than fluctuations below the average will decrease it. Thus, for a given average availability of resource, $\bar{R}$, the growth rate of species 1 will be unaffected by variance

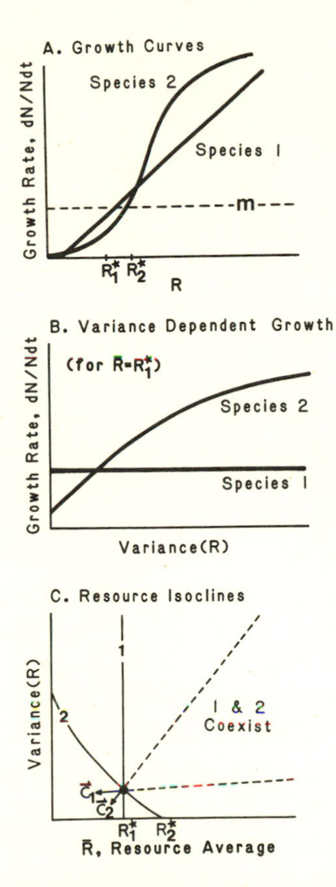

FIGURE 89. A. Resource-dependent growth curves for two species. Species 1 responds to the resources in a linear manner and species 2 responds to the resources in a sigmoid manner.

B. The curves of part A lead to these predicted dependences of the growth rate of each species on the variance in resource availability, for a situation in which the average resource availability is constant and equal to R_1^*. Because of its linear growth curve, the growth rate of species 1 is unaffected by variance, while the growth rate of species 2 increases with variance because of the approximately exponential shape of its growth curve in the region of $R = R_1^*$.

C. Because the growth rate of species 2 increases with Var(R), Var(R) is a resource for species 2. Competition between these two species for R and Var(R) is illustrated using the isocline approach. Because Var(R) is functionally a resource, these two species are capable of coexisting on "one" resource. This example is modified from Levins (1979).

in R, whereas the growth rate of species 2 will increase with variance, as shown in Figure 89B.

The increase in the growth rate of species 2 with increased variance in R means that variance in R (Var(R)) is a resource for species 2. It is thus possible to draw resource-dependent growth isoclines for these two species in relation to the two resources, R and Var(R). These are shown in Figure 89C. The isocline for species 2 shows that R and Var(R) are substitutable (complementary) resources for species 2. The point at which the ZNGI's of species 1 and 2 cross is a two-species equilibrium point at which the two species can potentially coexist. At this two-species equilibrium point, species 1 is limited by R and species 2 is relatively more limited by Var(R). The two species will stably coexist if species 1 "consumes" relatively more R than Var(R) compared to species 2. Species 1, with its linear growth function, consumes much less of the variance in R than does species 2. Thus, given the consumption vectors shown, the two-species equilibrium point is locally stable. This illustrates that a non-linearity in the growth function of a species can lead to the variance in the supply of a resource being responded to as if it were itself a resource, allowing the stable coexistence of two species.

Armstrong and McGehee (1980) illustrated another case of two species coexisting on one resource under non-equilibrium conditions. Figure 90A shows the resource-dependent growth curves of the two species for the limiting resource, R. Under equilibrium conditions, species 1 would displace species 2 because of its lower R^*. However, the interactions between species 1 and R in the absence of species 2 are locally unstable, as illustrated by the isoclines of Figure 90B. Thus, species 1 will cause oscillations in R, with N_1 and R oscillating around the locally unstable equilibrium point labelled X in Figure 90B.

How do these resource oscillations influence the average growth rates of species 1 and 2? Because of its linear growth function, such oscillations have no effect on the growth rate of

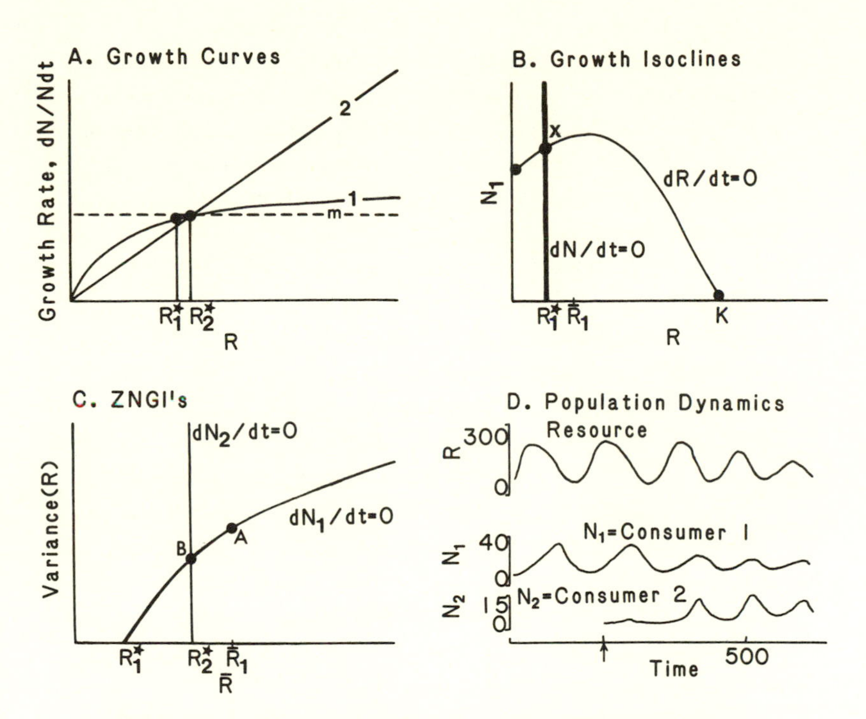

FIGURE 90. A. Resource-dependent growth curves for two species.

B. Lotka-Volterra predator-prey type isoclines for the interaction between species 1 and the resource, showing a locally unstable equilibrium point when this species consumes R. This leads to resource fluctuations.

C. The resource-dependent growth curves of part A lead to the ZNGI's shown. The shape of the ZNGI of species 1 indicates inhibition by increasing resource variance. Var(R) is thus a "limiting factor" for species 1. Such isoclines could be used to demonstrate the effect of any inhibitory substance, such as an alleopathic compound, on growth.

D. The results of a numerical simulation of competition between these two species for this single limiting resource. Note that population densities and R oscillate, but that the species still stably persist. This figure is redrawn from that of Armstrong and McGehee (1980).

species 2. Its long-term average growth rate is unaffected by the resource fluctuations. However, the fluctuations induced by species 1 mean that species 1 will require *more* of the resource on average to grow with an average rate equal to its mortality rate than if there were no fluctuations. This occurs because, with its Monod-shaped growth curve, the growth rate of species 1 decreases with increasing resource variance. Resource variance is thus a limiting factor for species 1.

This leads to the ZNGI's illustrated in Figure 90C. The straight-line ZNGI for species 2 means that its growth rate is unaffected by Var(R), while the curved ZNGI for species 1 illustrates that its growth rate is inhibited by Var(R). In the absence of species 2, the interactions between species 1 and R will lead to the point on the ZNGI of species 1 labeled A. At this point the average resource availability is $\bar{R}_1$, which is greater than R_1^*, and there is a certain Var(R) caused by the locally unstable interaction between species 1 and R. The addition of species 2 leads to a decrease in the average availability of R to a level of $\bar{R} = R_2^*$. This decrease in the average availability of R leads to a decrease in the average density of species 1, and thus a decrease in the Var(R). The two-species system moves toward the stable two-species equilibrium point B. Figure 90D shows a computer solution to the differential equations governing this interaction (Armstrong and McGehee, 1980). The oscillatory interactions between R and N_1 are shown in the absence of N_2 until day 194 (at the arrow) when N_2 was added. At that point the magnitude of the oscillations decreased for both R and N_1, and N_2 increased. N_1 and N_2 stably persisted, but not at fixed population densities, on R. Armstrong and McGehee (1980) further demonstrated that it was possible for an unlimited number of species to persist on a single limiting resource.

The assertion of Volterra (1931), MacArthur and Levins (1964), and others that no more species can stably coexist in a homogeneous, equilibrium habitat than there are limiting resources comes directly from all simple models of resource competition (such as Eq. 2). The vast discrepancy between this generalization and the diversity of natural communities led Hutchinson (1959, 1961) to pose the "paradox of diversity." Each of Hutchinson's suggested modifications of a simple model of resource competition seemingly can provide a solution to the paradox. The addition of spatial structure to a model of competition for two resources leads to a prediction of the long-term, equilibrium coexistence of an unlimited number of species (Chapter 5). Similarly, adding a trophic level to a model of resource competition predicts that an unlimited number of species can coexist, at equilibrium, on a single limiting resource. Also, the relaxation of the assumption of equilibrium coexistence to an assumption of long-term persistence, but not at fixed population densities, leads to the prediction that an unlimited number of species can persist in a homogeneous habitat on one or a few limiting resources. There are thus three different hypotheses, each of which has the potential to explain the diversity of natural communities. How may these competing hypotheses be tested and distinguished?

Ockham's Razor (Hutchinson, 1978) is a central tool of scientists in paring down the number of competing hypotheses to be tested. However, none of these three hypotheses can be rejected out-of-hand because of a higher level of complexity: each of these hypotheses results from the change of a single assumption in a simple model of resource competition. Similarly, none of the hypotheses can be rejected using observations on relationships between community diversity and the number of limiting resources, because each hypothesis can potentially explain the existence of any number of species on a given number of limiting resources. These hypotheses do differ, though, in their predictions about other aspects of community structure. These differences could be used to distinguish among them. The spatial heterogeneity theory suggests that species should be separated in a habitat in response to microsite to microsite differences in resource supply rates (see Chapter 6). The three-trophic-level model suggests that the species coexisting in a community should be inversely related in their competitive ability for a resource and their susceptibility to predation (Levin, Stewart, and Chao, 1977). The non-equilibrium model predicts sustained oscillations in the population densities of various species and in the availabilities of resources. Only much further experimental work will determine the relative importance of these competing hypotheses in explaining the observed diversity of natural communities.

PHENOTYPIC VARIABILITY

The theory presented in the preceding sections of this book has assumed that all the individuals within a species are identical in their resource requirements. What would happen if this assumption were relaxed? Would intraspecific phenotypic variability qualitatively change the predictions made earlier about the dependence of species composition and species richness on resource availabilities? How may phenotypic variability influence the interactions among several species?

Of all the phenotypic variability within a species, those phenotypes which are superior competitors for some habitats (i.e., for some resource supply points) are the phenotypes of greatest importance when considering interspecific competition among phenotypically heterogeneous species. Figure 91A illustrates what might be a random sample of phenotypes within a particular plant species consuming two essential resources. The kink points of those ZNGI's which are superior competitors for some resource supply points are accentuated with a large dot. These dots are connected with a smooth curve in Figure 91B. This curve may be called a phenotype tradeoff curve, for it shows the possible phenotypic tradeoffs for competitive ability for two limiting resources within a species. In a complete treatment of phenotypic variability, each point on

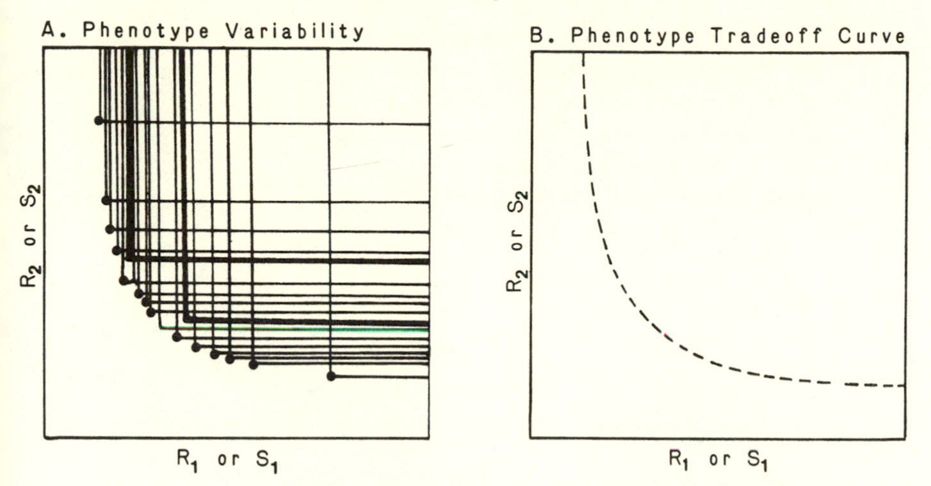

FIGURE 91. A. Phenotypic variability within a species. The ZNGI's of phenotypes which are superior competitors in some habitats are emphasized with a dot at the kink point of the ZNGI. Two ZNGI's for phenotypes which are inferior competitors in all habitats compared to the other phenotypes are shown with a heavier line. Also see Doyle (1975).
B. Phenotype Tradeoff Curve. A smooth curve may be used to join the kink points of the ZNGI's of the superior phenotypes. Such a curve shows all the viable tradeoffs in competitive abilities for R_1 and R_2 which can occur in this species.

such a tradeoff curve must be associated with a particular slope and length for the consumption vector. For this analysis, I assume that the tradeoffs involved are such that each phenotype consumes the resources in the proportion at which it is equally limited by the resources, and that the lengths of the consumption vectors are the same for all points on the curve.

We can then ask what happens when several phenotypically variable species compete for two resources. This is illustrated in Figure 92A. The broken line curves are tradeoff curves for two different species. The point at which the curves cross is a point at which the two species have identical ZNGI's. Resource supply points in the region labeled "*A* wins" will lead to dominance by various phenotypes of species *A*. Similarly, those in the region labeled "*B* wins" will lead to dominance

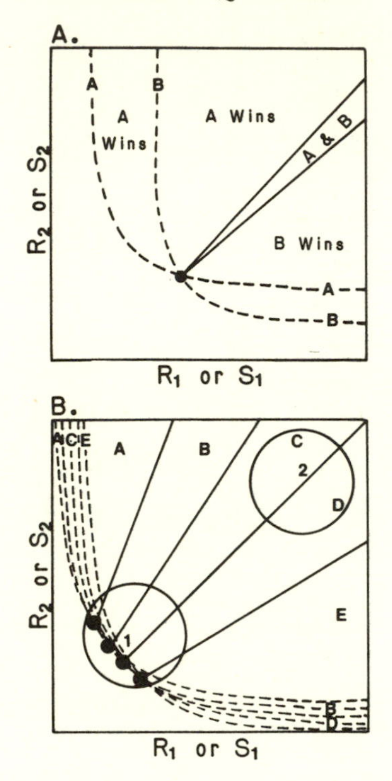

FIGURE 92. A. Competition between two phenotypically variable species. The broken lines are phenotype tradeoff curves, not ZNGI's.

B. Competition among five phenotypically variable species in a heterogeneous habitat, indicated by the circles. Note that species richness decreases with increased resource richness.

by various phenotypes of species *B*. There will either be a region of coexistence coming from the equilibrium point (as drawn), or there will be a straight line coming from this point along which both species may coexist. The region of coexistence will become increasingly thin, approaching a straight line, as the number of phenotypes along the tradeoff line increases. (The region of coexistence is shown as a line in future figures.) As for cases considered earlier in which phenotypically homogeneous species were competing, spatial heterogeneity of resources can allow the equilibrium coexistence of more species than there are limiting resources (Fig. 92B). Figure 92B also illustrates that species diversity should depend on resource richness, as previously discussed, even when the assumption of genetic homogeneity is relaxed. A moderately resource-poor habitat, such as habitat 1, includes microsites in which all five species may coexist. Enrichment of this habitat to habitat 2 would decrease the community species richness to two species. Further enrichment could lead to a single-species community.

This simple consideration of phenotypic variability suggests that the addition of phenotypic variability to the resource-based model developed in this book will not change the qualitative, community-level predictions made. Species richness would still have the same predicted dependence on resource richness, and, as inspection of Figure 92B reveals, species composition of habitats would still be predicted to depend on the relative rates of resource supply (resource ratios) in the habitats.

THE SUPERSPECIES PROBLEM

In addition to assuming phenotypic homogeneity within a species, the theory presented in the preceding chapters has assumed tradeoffs in the resource competitive abilities of different species. A species which is a superior competitor for one resource has been assumed necessarily to be an inferior competitor for another resource. Similarly, the brief discussion of phenotypic variability in the preceding pages has assumed tradeoffs within a species such that individuals which have a phenotype which makes them a superior competitor for one resource necessarily are inferior competitors for another resource. There is an alternative assumption, the assumption of a superspecies.[1]

[1] The problem of the evolution of a superspecies was brought to my attention by a comment made by G. Evelyn Hutchinson, relayed to me by Peter and Susan Kilham.

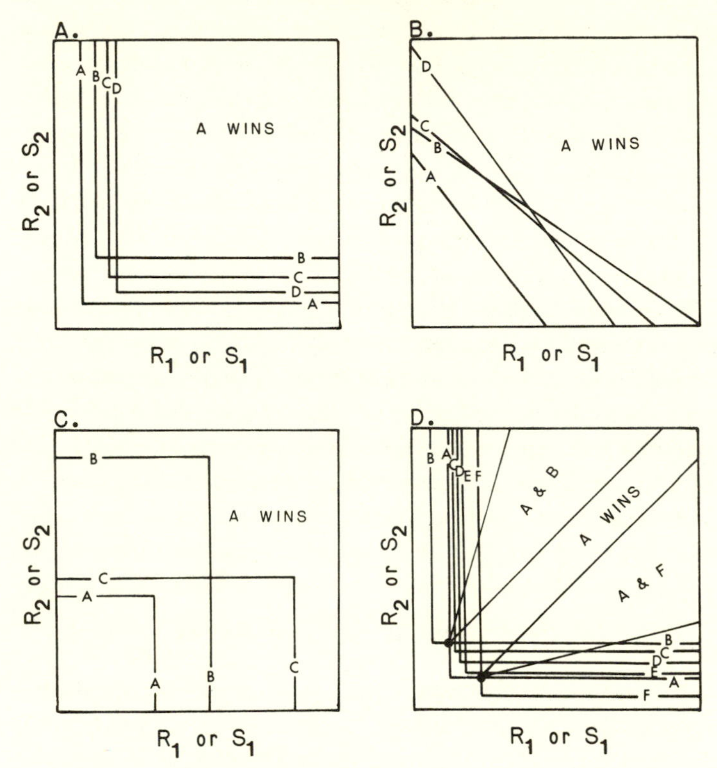

FIGURE 93. A, B, and C. Species *A* is a superspecies compared to the other species of these figures. It would competitively displace all other species from any habitat in which it or they could survive.

D. Species *A* is a partial superspecies in this case. It would displace species *C*, *D*, and *E*, but coexist with species *B* and *F*.

Figure 93 illustrates several possible cases of superspecies. For parts A, B and C of Figure 93, species *A* is a superspecies, a superior competitor for both limiting resources compared to all other species. In these three cases, the ZNGI of species *A* is inside that of all other species. Species *A* is predicted to competitively displace all other species no matter how spatially heterogeneous the habitat may be. For Figure 93D, species *A* is a "partial" superspecies. If individuals of species A invaded a community composed of species B to F, it would displace three species, reducing species richness by 50%. These examples suggest that the community-level patterns discussed in the preceding chapters depend on the assumption that there is no superspecies, since the appearance of a superspecies would lead to a very species-poor community. If superspecies do sporadically arise, how can communities with high diversity continue to exist?

An initial answer to this question might seem to come from a comparison of parts A, B and C with part D of Figure 93. A superspecies which is a superior competitor compared to only a few species (a partial superspecies, such as species A in Fig. 93D) would seem more likely to arise than a superspecies which is a superior competitor compared to all other species. The evolution of an occasional partial superspecies may not seem to be a great problem, for it would result in the displacement of only a few species. But the long-term effect of the evolution of many such partial superspecies would be a continual decrease in species richness, a trend not supported by the fossil record.

Selection within a Superspecies

What would happen if a superspecies arose which was able to competitively displace all other species requiring the same set of resouces? Let us consider two situations—a superspecies for two essential resources (as in Fig. 93A) and a superspecies for two switching resources (as in Fig. 93C). Let us consider first a superspecies for essential resources and let us assume that the species is initially phenotypically homogeneous and lives in a spatially homogeneous habitat. The resource supply point for this habitat is labeled A, and the ZNGI for this species is labeled 1 (Fig. 94A). For supply point A, all the individuals of this species are limited by R_1. What if there were phenotypic variability? Any individual within this species which was a

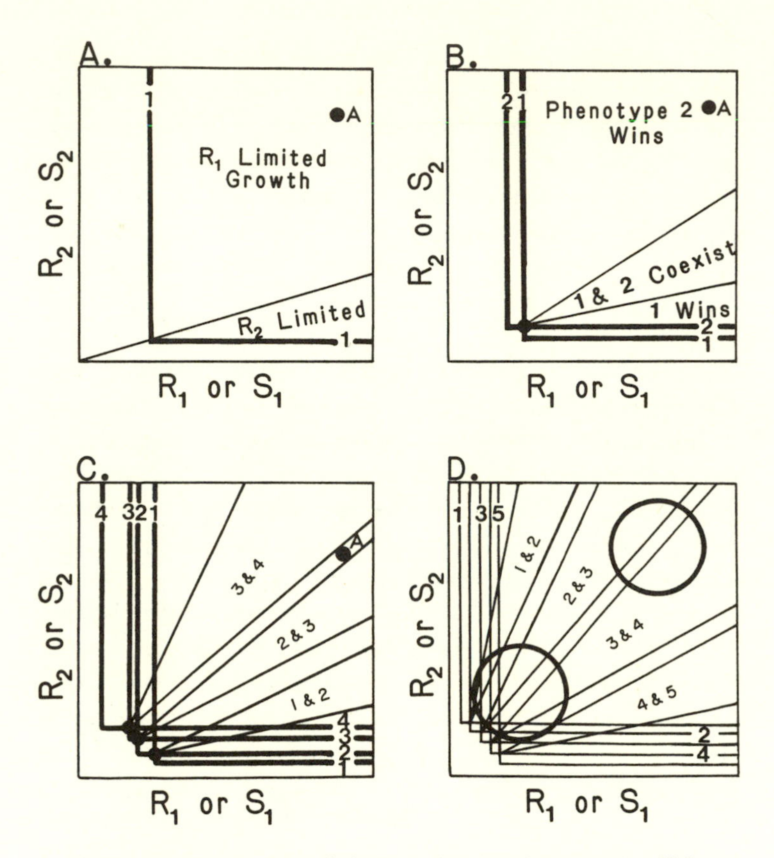

FIGURE 94. A. In habitat *A*, individuals with phenotype 1 will be limited by R_1.

B. Individuals which are superior competitors for R_1—even if to be so requires that they be inferior competitors for R_2—will be favored in habitat *A*. Thus, phenotype 2 will competitively displace phenotype 1 from habitat *A*.

C. Of all the phenotypes that may arise, individuals with a phenotype that has its ZNGI nearest the origin and leads to the individuals being equally limited by both resources will be favored. Phenotype 3 is such a superior phenotype for habitat *A*. Individuals with phenotypes 1, 2, and 4 will be at a competitive disadvantage compared to those with phenotype 3 in habitat *A*.

D. Spatial heterogeneity, indicated by the circles, can lead to the coexistence of many phenotypes, with more phenotypic variability in low resource habitats.

superior competitor for R_1, even if this meant that it was an inferior competitor for R_2, would be at a selective advantage. Thus, if it were possible to gain competitive ability for R_1 by reducing an individual's competitive ability for R_2, such an individual would be favored.

One such case is shown in Figure 94B. An individual with phenotype 2 is a superior competitor for R_1 but an inferior competitor for R_2 compared to individuals of phenotype 1. Individuals of phenotype 2 would outreproduce and eventually competitively displace individuals with the original phenotype. Similarly, an individual with phenotype 3 of Figure 94C is a superior competitor for R_1 but an inferior competitor for R_2 compared to phenotype 2. Individuals of phenotype 3 would displace individuals with any of the other three phenotypes shown. This process would continue as long as phenotypes arose which were superior competitors. The optimal phenotype would be the phenotype with its ZNGI nearest the origin and which consumed the two limiting resources in the proportion in which they were supplied. This would lead to equilibrium environmental availabilites of the resources at the "kink point" of the ZNGI. Individuals with such a phenotype would be at a competitive advantage in habitat *A* compared to individuals which have tradeoffs which make them better competitors for either R_1 or R_2. In a spatially homogeneous environment, selection for competitive ability for essential resources within a superspecies should generate a phenotypically homogeneous population which is equally limited by all resources. In spatially heterogeneous habitats, phenotypic diversity would be favored, as illustrated in Figure 94D.

Let us next consider a superspecies for two switching resources. Let us start by assuming that the newly arisen superspecies is phenotypically homogeneous with the ZNGI labeled 1 and that the habitat is spatially homogeneous, with its supply point being point *A* of Figure 95A. What would happen if some phenotypically different individuals arose? Phenotype 2 of

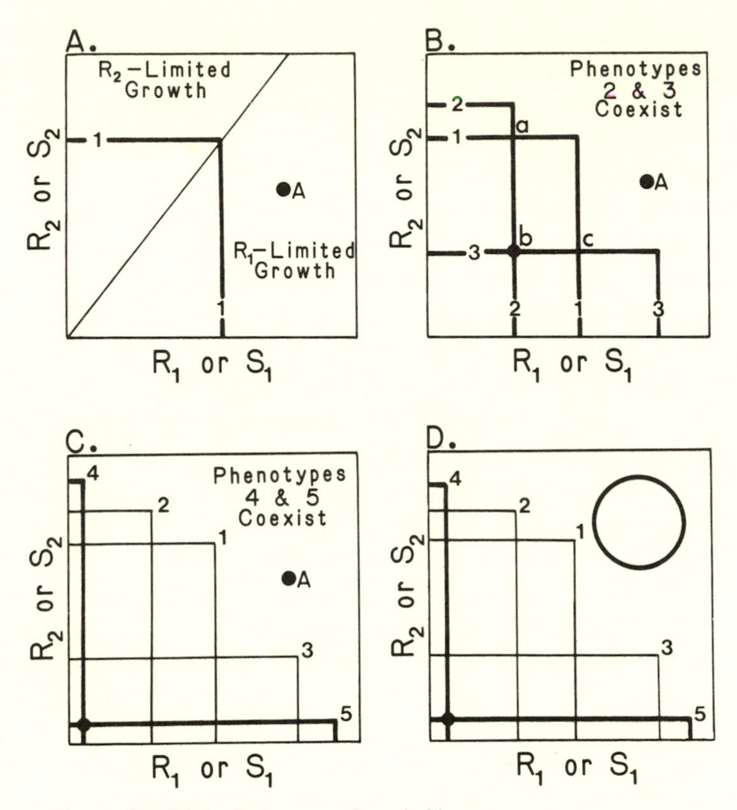

FIGURE 95. Optimal phenotype for switching resources.

A. These individuals are limited by the availability of R_1.

B. Individuals of phenotype 1 will be competitively displaced by individuals of phenotypes 2 and 3, where phenotype 2 is a superior competitor for R_1 (but inferior for R_2) and phenotype 3 is a superior competitor for R_2 but an inferior competitor for R_1. Thus, the original phenotype is displaced by individuals which are increasingly specialized on either one or the other resource.

C. Phenotypes 1, 2, and 3 would be displaced by phenotypes 4 and 5, which are even more specialized on R_1 and R_2. With switching resources, selection will favor those phenotypes completely specialized on one resource even if they are incapable of using the other resource.

D. Spatial heterogeneity, indicated by the circle, only allows the coexistence of two phenotypes—those specialized on one or the other resource.

Figure 95B is a superior competitor for R_1 but an inferior competitor for R_2 compared to phenotype 1, while phenotype 3 is a superior competitor for R_2 and an inferior competitor for R_1 compared to phenotypes 1 and 2. There are three points of intersection of these ZNGI, labeled *a*, *b*, and *c*. Only the equilibrium between phenotypes 2 and 3, equilibrium point *b*, is locally stable. The equilibrium point where the ZNGI's of phenotypes 1 and 2 cross (point *a*) is not stable because it is outside the ZNGI of phenotype 3. Phenotype 3 is able to continue growing for resource levels at point *a*, and it will reduce R_2 down to a point on its ZNGI. Similarly, the equilibrium point at which the ZNGI's of phenotypes 1 and 3 cross (point *c*) is unstable because phenotype 2 can reduce R_1 down to a point on its ZNGI. This means that phenotypes 2 and 3 will drive phenotype 1 to extinction. Thus, new phenotypes which are increasingly specialized on either R_1 or R_2 compared to the original phenotype will be favored, and the original phenotype will be competitively displaced by the two variants. This process can continue as long as variants arise which are superior competitors for one or the other resource, as illustrated in Figure 95C. Here phenotypes 4 and 5 will competitively displace all other phenotypes. These graphs illustrate, for a species consuming switching resources in a homogeneous habitat, that selection will favor those phenotypes which specialize on one or the other resource. Eventually, for a homogeneous habitat in which both resources are being supplied at a constant rate, a superspecies consuming two switching resources will have two phenotypes: individuals specialized on R_1 (and almost incapable of living on R_2), and individuals specialized on R_2 (and almost incapable of living on R_1). Figure 95D illustrates that this same conclusion is reached when a spatially heterogeneous habitat is considered. A switching species in a heterogeneous habitat will have only two optimal phenotypes, one specialized on one resource and the other specialized on the other resource.

The ultimate tradeoff in competitive ability for one versus the other switching resource would be for an individual to lose its ability to consume the resource on which it did not specialize. Such an individual would be completely dependent on a single resource (host, food plant, etc.). In a perfectly predictable habitat, such a tradeoff would be selectively favored if it were to lead to increased competitive ability for the favored resource. However, such extreme specialization would not be favored by natural selection if the resource upon which the individual was specializing fluctuated in availability, especially if there were times that it was rare or absent. Temporal fluctuations in resource availability should limit the degree of specialization for a species consuming switching resources.

Superspecies and Limits to Diversity

How, then, may the evolution of a superspecies and selection on individuals within a superspecies influence the diversity of a community? Consider a recently evolved superspecies, a plant which is consuming two essential resources in a spatially heterogeneous habitat. As the population size of this plant increases and the inferior species are competitively displaced, selection acts within this superspecies to favor phenotypes which are superior competitors for their particular microhabitats (Fig. 94D). Once the superspecies occupies the full range of the spatially heterogeneous habitat, a wide variety of phenotypes will be favored in the various parts of the range of the species (Fig. 94D). Because major habitat types are likely to occur as physiographically distinct units, the microsites within a local habitat should be relatively similar and thus should contain fairly similar phenotypes. Because nearby individuals are more likely to breed with each other, the broad range of phenotypes within the superspecies would tend to occur as sets of (at least partially) spatially distinct subpopulations. Given such spatial separation of phenotypically different subpopulations experiencing different selective pressures, speciation is likely to occur. Speciation would also be favored by any processes that led local populations to have different times or modes of reproduction. Thus, although the evolution of a superspecies would initially reduce the species richness of a community by excluding species, speciation within the superspecies would eventually lead to a group of species derived from the original superspecies, with each new species specialized on a particular type of microhabitat, i.e., a particular resource ratio. Such patterns of specialization within the superspecies depend on the assumption of tradeoffs within the superspecies in competitive abilities for the various limiting resources. These tradeoffs within a species can lead to a multispecies-complex derived from the ancestral superspecies.

Patterns within numerous plant communities are suggestive of such a process. Many plant communities are dominated by one or a few taxonomic groups. Each taxonomic group may be descended from a superspecies which was a superior competitor for the particular physical conditions and resource limitations of the area, with speciation occurring within the taxonomic group in response to microsite differences. The eucalypts of Australia comprise a group of over 700 species derived from a few ancestral species which invaded the nutrient-poor soils of the Australian continent. Numerous studies suggest that these species are specialized on different soil types. Similarly, the fynbos vegetation of the Cape of South Africa, the most diverse vegetation in the world (Kruger and Taylor, 1979), is located on very nutrient-poor soils. Five different plant families are represented by more than 100 species each in one site studied, and 8 families are represented by over 100 species each throughout the entire Cape region. For many of these families, the species present represent a group derived from what was probably a single ancestral superspecies. The very diverse forests of Malaysia are composed mostly of species in the Dipterocarpaceae (Poore, 1968), and may represent a species complex derived from a superspecies ancestor. For mid-latitude, mildly productive lakes, the majority of the species are diatoms in the Fragilariaceae, again suggesting a superspecies complex which was specialized on a particular pattern of nutrient availabilities (conditions of low phosphate and moderate to low levels of silicate). The process of speciation within a superspecies complex suggests that the evolution of a superspecies need not lead to an ultimate decrease in diversity.

Indeed, for plants competing for essential resources in a spatially heterogeneous environment, quite a different problem seems to exist. There does not appear to be any simple limit placed on the number of species that can invade and stably coexist in a spatially heterogeneous environment. This is illustrated in Figure 96. A plant community composed of species *A* and *B* can be invaded, in theory, by species which are intermediate in their resource requirements, such as species *D*, and by species which have a tradeoff which makes them better competitors for either R_1 or R_2, such as species *C* and *E*. This community of five species can be invaded by any other species with intermediate requirements for R_1 and R_2, without limit. Whatever the origin of invading species—be they newly evolved species in a superspecies complex or propagules from a species already in existence in a different location—there is no simple limit on the number of such species that can coexist within a plant community with any given level of spatial heterogeneity. This suggests that the question of the evolution of species diversity within communities of plants is much more complex than the question of the factors that can maintain diversity in a community with an existing species pool.

The question of the limits to the number of species which may exist in a community, given a sufficiently long period of time for the processes of speciation and extinction to go to an equilibrium, is really a question of the limits to similarity within a community. This question has been addressed from different perspectives by several workers (e.g., Hutchinson, 1959; MacArthur and Levins, 1967; May and MacArthur, 1972;

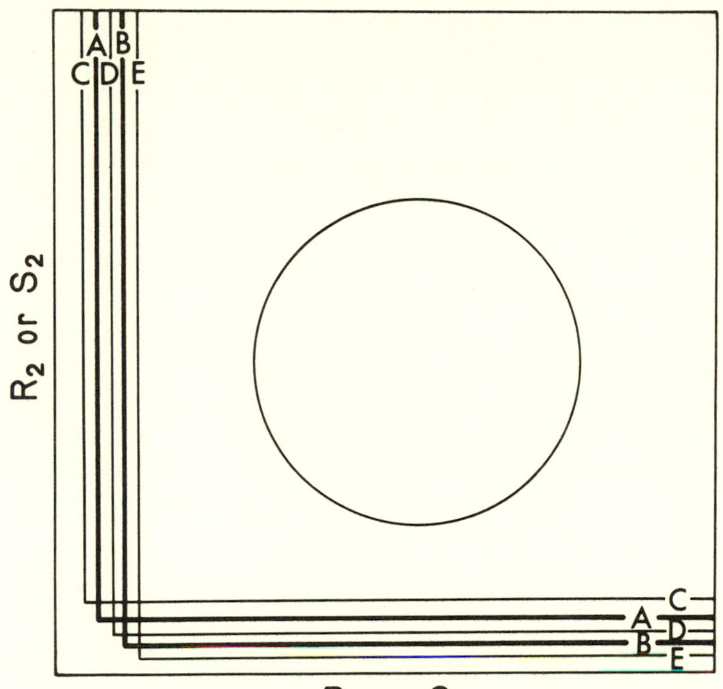

FIGURE 96. The community composed of species *A* and *B* could be invaded by species *C*, *D*, and *E*, giving a community of five species in this heterogeneous habitat, indicated by the circle. This community could in turn be invaded by an unlimited number of other species which had intermediate requirements for R_1 and R_2 compared to any adjacent pairs of species already existing in the community. In theory, there is no simple limit to the number of species that can invade and coexist on two essential resources in a spatially heterogeneous habitat.

May, 1973; Abrams, 1975; Turelli, 1978), and will not be reexplored here. However, I would like to mention briefly some of the factors that might be involved in establishing an equilibrial species richness in a community of plants competing for limiting resources.

An equilibrium will occur when, on average, each successful invasion by or evolution of a new species leads to the extinction of a species present in the community and when the extinction of a species present leads to the eventual successful invasion by or evolution of another species. The major factor favoring increased species richness is the potential ability of any species with an intermediate resource requirement to invade and stably coexist in a plant community (Fig. 96). However, the chance that a species will successfully invade should decrease with species richness because the more species there already are in a community, the less likely it is that a propagule of an invading species will colonize a microsite in which it is a superior competitor. This occurs because the range of microsites for which a species is a successful competitor decreases as the species richness of the community increases. The more similar an invading species is to the species already present, the smaller will be the range of microsites which it can successfully invade. This suggests that the rate of addition of new species to a community should decrease as species richness increases, but it does not provide a mechanism which places a strict limit on diversity. Such a limit must come from extinction rate increasing with species richness. There are several reasons to suppose that this occurs. First, the greater the number of species within a community, the lower will be the population size of each species. As the population size of a species becomes smaller, the total number of propagules produced by the population decreases. However, as there are more species in a community the number of microsites in which these propagules will be successful competitors also decreases. This suggests that, for any species, there will be a minimum population size below which extinction would occur because of a low probability that propagules would colonize the proper microsites. A similar argument has been made by Levington (1979), in which he suggested that random processes would impose a minimum population size on each species, and that this would limit diversity. In terms of species competing for essential resources, the invasion of a species will tend to reduce the population sizes of the species most similar to it, and thus should lead to either their extinction or its extinction. Another process which could limit diversity is the evolution of superspecies. Although speciation within a superspecies can lead to increases in diversity, the evolution of new superspecies or partial superspecies would decrease diversity. These various processes can establish an upper limit to the diversity of a community. However, the relationship between such a limit on the evolutionarily possible diversity of a community and the short-term limits discussed earlier has not yet been explored.

WHY ARE THERE SO FEW ANIMAL SPECIES?

Grubb (1977) asserted that the existence of "a million or so animals can be easily explained in terms of the 300,000 species of plants (so many of which have markedly different parts such as leaves, bark, wood, roots, etc.), and the existence of three to four tiers of carnivores (Hutchinson 1959)," but "there is no comparable explanation for autotrophic plants; they all need light, carbon dioxide, water and the same mineral nutrients." This assertion might make one believe that this section should be titled "Why Are There So Many Plant Species?," and on its surface does not seem to raise any questions about the diversity of animal communities. However, let us turn Grubb's statement around. The experimental work done on natural plant communities has revealed that there are two to five resources which limit the growth of plants in any one area. Worldwide, there are no more than 30 resources that limit the growth of plants, for 30 resources would include light, several different forms of open space, carbon dioxide, and all the mineral elements that plants require. Thus, with at most 30 limiting resources, it seems that about 300,000 species of terrestrial plants have been able to evolve and coexist worldwide. This suggests that the processes controlling worldwide plant diversity are such that the ratio of plants to their limiting resources is at least 300,000 to 30, or 10,000:1. This ratio may be compared to that for animal communities. Following Grubb's suggestion that each plant species is more than one resource to an animal ("leaves, bark, wood, roots, etc."), we might hazard a guess that the 300,000 species of plants represent about a million distinct animal resources. This suggests that the ratio of animal species to their resources is approximately 1,000,000 to 1,000,000 or about 1:1. If the process of local spatial heterogeneity is the main mechanism allowing the coexistence of many more plant species than limiting resources, as the discussion of Chapter 5 suggests, we might expect that this same process would allow the coexistence of a comparable ratio of animal species to their limiting resources. If this were the case, we would expect to have about 10^{10} species of animals, i.e., about 10^4 times the number actually observed. Either the National Science Foundation has been too conservative in funding basic taxonomic research, or there is a major difference between the processes that determine the diversity of plant and animal communities.

I have been perplexed by this apparently major difference between plant and animal diversity ever since I first turned Grubb's assertion on its head. I offer two speculative "solutions" to this paradox in hopes that they may stimulate interest in a question which I find intriguing. First, it may be that plant community diversity is not determined by resource spatial heterogeneity as suggested in this book, but comes from trophic complexity. If plant community diversity were caused by trophic complexity, the number of plant species should be less than or equal to the sum of the number of limiting resources and the number of herbivore species (as discussed in a preceding section of this chapter). There would then be an approximately 1:1 ratio between species and their "resources" for both plant and animal communities.

But what of the alternative possibility? If resource spatial heterogeneity does determine plant diversity, why does it not similarly affect animal communities? One factor might be the difference between the type of resources consumed by plants and motile animals. As already discussed, most of the resources consumed by plants are essential. However, most of the resources consumed by motile animals are nutritionally substitutable for each other.

As discussed in Chapter 2, nutritionally substitutable resources will often be consumed in a switching manner by motile organisms. When a consumer switches between two resources, it consumes the one resource which currently leads to the greater fitness, switching to the other resource when the availabilities of the resources have changed so that the other resource leads to greater fitness. Switching is an optimal foraging tactic in spatially homogeneous habitats (Fig. 4A) and in spatially heterogeneous habitats in which local availabilities of two resources are inversely correlated (Fig. 4C). Nutritionally substitutable resources should be consumed in a non-switching manner only if the resources occur together in local patches and can be found and consumed using similar behaviors. Thus, switching growth isoclines might be considered the most general response of a motile animal to non-complementary resources.

Previous sections of this chapter have explored some possible effects of phenotypic variability for species with essential and switching resources. This analysis suggested that selection within a species with essential resources could lead to the coexistence of many different phenotypes in a spatially heterogeneous habitat. Each of the phenotypes would be specialized on a particular ratio of availabilities of the two essential resources. In contrast, selection within a species with switching resources would favor individuals which were increasingly specialized on one or the other resource. At equilibrium, no more than two phenotypes could coexist within a species limited by two switching resources, no matter how spatially heterogeneous the habitat.

A comparable situation arises when several switching species compete for two resources, as illustrated in Figure 97A. Of the

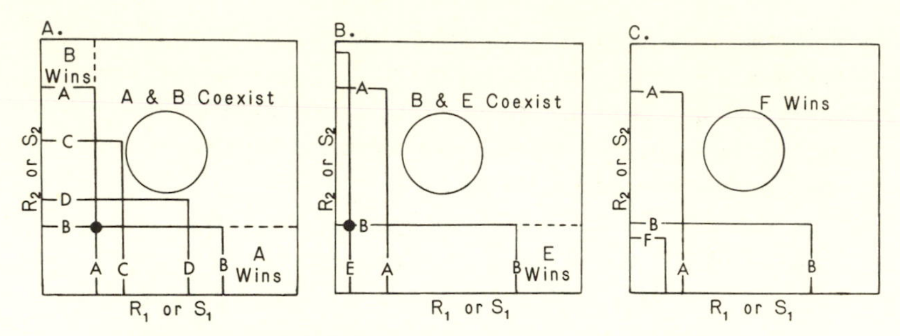

FIGURE 97. A. In competition among any number of species for switching resources, only two species can coexist no matter how spatially heterogeneous the habitat may be.

B and C. The only species capable of invading such a system must be better at one or the other resource, and will necessarily displace one (part B) or both species (part C).

four species present only two will be able to coexist no matter how spatially heterogeneous the environment may be. These are species *A* and species *B*. The other species, which are intermediate in their requirements for these two switching resources, are competitively displaced. A community of species *A* and *B* could be invaded only by a species which is a better competitor for one or both of the switching resources. Species *E* of Figure 97B can invade this community because it is a superior competitor for R_1, but it will displace species *A* when it does invade. Similarly, species *F*, which is a superior competitor for both resources, can invade the community, but it will displace both species *A* and *D* in so doing. No matter how spatially heterogeneous the habitat, there can be no more switching species than there are resources. Thus, if selection for optimal diet of motile animal species means that most animals will have switching resources, there should be an approximately 1:1 relationship between the number of animal species and the number of their resource species.

To test the robustness of this statement, let us explore two possible exceptions. The first is the possibility of invasion of a community of switching species by a non-switcher. Figure 98A

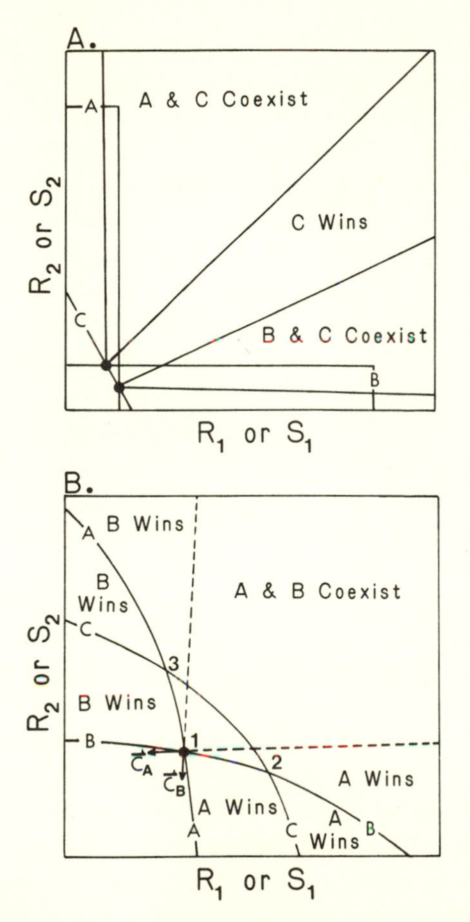

FIGURE 98. A. Only non-switching species can invade a community of switching species and lead to increased species richness. However, a non-switcher must be a superior competitor compared to both of the switching species to do this.

B. The 1:1 relationship between species richness and number of resources does not depend on perfect switching, as illustrated above. Species *C* will be competitively displaced by species *A* and *B* under all conditions.

illustrates such a possibility. In this case, species *C* is capable of invading the community because its response to these two resources gives it a straight-line isocline (i.e., the resources are perfectly substitutable for species *C*). The addition of this non-switcher gives two two-species equilibrium points. This would then allow the coexistence of more species than there are resources, and could let the diversity of animal communities depend on spatial heterogeneity as hypothesized for plant communities in Chapter 5. However, I suggest that a species such as species *C* will rarely occur. In order to invade the community, there must be a range of availabilities of R_1 and R_2 for which species *C* has a lower equilibrium requirement for R_1 than does species *A* and a lower equilibrium requirement for R_2 than does species *B*. Species *A*, however, exists in the community because it is the most efficient consumer of R_1, just as species *B* exists there because it is the most efficient consumer of R_2. Thus, for a non-switcher to invade a community of switching species, it must be a superior competitor compared to species which are specialists on each resource. The non-switching "generalist" can thus invade only if there is a range of resources for which it wins over both specialists. With the selective forces that act within a switching species, it seems unlikely that there will be many non-switchers capable of invading such a community of switching species.

The second possible exception is potentially more important. The analysis presented in Figure 97 assumed that the species were perfect switchers, i.e., that their isoclines had a perfect right-angle corner. This seems unlikely for real organisms. Do the generalizations derived for perfectly switching species hold when species are not perfect switchers? Consider the three species of Figure 98B. The isoclines drawn there are for antagonistic resources. Species *A* is the best competitor for R_1 and species *B* is the best competitor for R_2. Species *C* is intermediate in its competitive abilities for R_1 and R_2. There are three points of intersection of these isoclines. Point 1 is a stable equilibrium for species *A* and *B*, for species *C* cannot survive at this resource level and species *A* and *B* consume more of the resource which more limits their own growth. However, point 2 is not an equilibrium point because point 2 is outside the ZNGI of species *A*. Similarly, point 3 is not an equilibrium point because it is outside the ZNGI of species *B*. Relaxation of the assumption that consumers are perfect switchers does not lead to additional two-species equilibrium points, and thus does not allow more species to coexist than there are resources.

This brief analysis suggests that at least part of the reason for the great discrepancy between the ratio of plant species to their resources compared to the ratio of animal species to their resources may come from the different type of resources consumed by plants and animals.

The essential resources consumed by plant species allow each plant species to specialize on a proportion or ratio of the essential resources. A species will tend to be dominant in communities that have supply rates of resources that fall near that optimal ratio. Thus, point to point heterogeneity in the supply process of essential resources can allow the coexistence of many more plant species than there are limiting resources by creating a series of microhabitats that differ in the ratios of limiting resources.

The mobility of animals, combined with the tendency for their resources to be nutritionally substitutable, means that many animal species will be switchers. Switching does not allow specialization on a particular proportion of resources. Thus, communities of switching consumers will not have greater species richness with increased spatial heterogeneity in the supply rates of the switching resources. There should be approximately a 1:1 ratio of animal species to their resources.

SUMMARY OF BOOK

The theoretical approach developed in this book uses information on the resource requirements of species to predict the equilibrium outcome of multispecies competition for several limiting resources. An extension of this graphical, equilibrium theory to multispecies competition in spatially heterogeneous habitats suggests that the diversity of plant communities should be maximal in moderately resource-poor habitats. Species diversity should decline with either increases or decreases in resource richness. Thus, the theory predicts that a graph of species richness against resource richness should give a humped curve. A brief extension of the theory developed for plant communities to communities of sessile, space-limited animals suggests that a similar pattern should occur in these communities. Just as nitrogen and phosphorus are essential resources for plants, so open space and food are essential resources for sessile animals. Theory predicts that sessile animal communities should have maximal diversity at moderately low supply rates of space—i.e., at moderate disturbance rates. For light-limited plant communities, disturbance increases the light available at the soil surface (and possibly other resources), and thus disturbance in light-limited plant communities should similarly influence diversity.

The theory presented in this book predicts that the species composition of a plant community should be determined by the relative availabilities of the limiting resources—i.e., by the ratios of the resource supply rates. Spatial heterogeneity in resource supply rates can be easily incorporated into the graphical models of competition if the heterogeneity is measured as the individual-to-individual variance in resource supply rates. Temporal variation in resource supply rate may allow species to respond to a single resource as if it were several resources or as if it were a resource and a limiting factor. Either spatial or temporal variations in resource supply rates may allow many more species to stably persist than there are limiting resources. The addition of a third trophic level may also allow many more species to coexist than there are limiting resources. Another complicating factor which is likely to have profound effects on resource competition, but which has not been considered in this book, is age structure.

The theory developed in this book also suggests that the structure of motile animal communities may differ greatly from that of plant (and sessile animal) communities because of the different types of resources required. Whereas plant communities and communities of sessile animals may have many more species than there are limiting resources, communities of motile animals should have their diversity limited to the number of resources available. The theory also suggests that plant community succession may be at least partially explained in terms of changes in the ratios of limiting resources.

In addition to these theoretical predictions, this book has reviewed a portion of the large data set which has accumulated on the relationship between resource availability and plant community structure. This review suggests (1) that the outcome of competition between a few plant species under controlled conditions can be predicted by a simple model of resource competition; (2) that fertilization of plant communities leads to decreased species diversity as predicted by theory; (3) that natural plant communities often show a humped relationship between species richness and resource richness; and (4) that individual plant species are dominant only through a narrow range of resource ratios.

In summarizing this work, I am drawn back to the thoughts that Hutchinson expressed in his "Homage to Santa Rosalia." Clearly, there is much more to communities than can be understood in terms of consumer-resource interactions. However, I believe that the simple, graphical approach to resource competition developed in this book, and its extensions to questions of community diversity, can become the cornerstone of a theory of the structure and functioning of natural communities. As has been repeatedly stressed in the recent ecological literature (e.g., Neill, 1975; Smith-Gill and Gill, 1978; Lynch, 1978) it is only through knowledge of the mechanisms of interactions of species that we can build a predictive theory of interspecific interactions. What are called mechanisms depends on the organisms considered. Nutrient-limited uptake and growth are mechanisms of plant competition, just as behavior and foraging tactics are methods of interaction for animal species. This book has outlined one approach that can be taken to add elements of mechanism to the study of populations and communities. I have concentrated on plant communities; perhaps similar approaches will be useful in studies of animals.

APPENDIX

Mathematics Associated with Chapter 7

Chapter 7 presented four cases of the analogies between the resource-approach to competition and the classical Lotka-Volterra approach to competition. This Appendix expands on the mathematics used to derive these relationships. Because the mathematical details are similar for all four cases, the first case will be presented in the most detail.

As stated in Chapter 7, at equilibrium ($dN_1/dt = dN_2/dt = 0$) the Lotka-Volterra competition equations simplify to the following equations:

$$N_1^* = K_1 - \alpha N_2^* \qquad \text{and} \tag{1}$$

$$N_2^* = K_2 - \beta N_1^*, \tag{2}$$

where N_i^* is the population density of species i at equilibrium when it is competing with other species; K_i is the carrying capacity of species i, i.e., the equilibrium population density of species 1 in the absence of interspecific competition; α is the competition coefficient which measures how each individual of species 2 depresses the potential carrying capacity of species 1; and β is the competition coefficient which measures how each individual of species 1 reduces the potential carrying capacity of species 2.

In order to derive the analogs of the carrying capacities and the competition coefficients from a given model of resource competition, it is necessary to solve the model for equilibrium and then rework the equations until they are in the form shown above.

Let us first consider a case of two species competing for perfectly substitutable resources supplied in an equable mode. This is the first case of Chapter 7. Let us use the Monod model of competition to model the dynamics of resource competition for these two resources. The Monod model would have the following form for perfectly substitutable resources:

$$dN_1/N_1\ dt = r_1 \frac{w_{11}R_1 + w_{12}R_2 - T_1}{k_1 + w_{11}R_1 + w_{12}R_2 - T_1} - D, \tag{3}$$

$$dN_2/N_2\ dt = r_2 \frac{w_{21}R_1 + w_{22}R_2 - T_2}{k_2 + w_{21}R_1 + w_{22}R_2 - T_2} - D, \tag{4}$$

$$dR_1/dt = D(S_1 - R_1) - c_{11}N_1 - c_{21}N_2, \tag{5}$$

$$dR_2/dt = D(S_2 - R_2) - c_{12}N_1 - c_{22}N_2, \tag{6}$$

where R_1 and R_2 are the availabilities of resources 1 and 2; r_1 and r_2 are the maximal reproductive rates of species 1 and 2; N_1 and N_2 are the population densities of species 1 and 2; D is the mortality rate experienced by both species and the supply rate constant for the limiting resources; k_1 and k_2 are the Monod model half saturation constants for species 1 and 2; T_1 and T_2 are the minimal amounts of the weighted values of R_1 and R_2 required for survival of species 1 and 2; w_{ij} is the weighting factor which converts the availability of R_j into its value for species i; and c_{ij} is the amount of R_j consumed per individual of species i.

When solved for equilibrium, Equations 3 and 4, which describe the growth of the two species, yield the resource-dependent growth isoclines for the species. These isoclines, graphed in Figure 73, are defined by the following equations:

$$R_2 = -w_{11}/w_{12}R_1 + B_1 \quad \text{(isocline for species 1)}, \tag{7}$$

$$R_2 = -w_{21}/w_{22}R_1 + B_2 \quad \text{(isocline for species 2)}, \tag{8}$$

where $B_1 = (D(k_1 - T_1) + r_1T_1)/(w_{11}(r_1 - D))$ and $B_2 = (w_{21}/w_{22})(D(k_2 - T_2) + r_2T_2)/(w_{21}(r_1 - D))$. When $dR_1/dt = dR_2/dt = 0$, Equations 5 and 6, for the dynamics of resource change, give the following relationships:

$$(\text{for } dR_1/dt=0) \quad N_1^* c_{11} + N_2^* c_{21} - D(S_1 - R_1^*) = 0, \quad (9)$$

$$(\text{for } dR_2/dt=0) \quad N_1^* c_{12} + N_2^* c_{22} - D(S_2 - R_2^*) = 0. \quad (10)$$

Equations 9 and 10 define the equilibrium population densities of the two species relative to the equilibrium availabilities of the resources. The equilibrium concentrations of the resources, R_1^* and R_2^*, are the concentrations at which the isoclines of Figure 73 cross, i.e., the concentrations of the two-species equilibrium point. These concentrations are obtained by solving Equations 7 and 8 of this Appendix. When these values for R_1^* and R_2^* are substituted into Equations 9 and 10, the following equations are obtained:

$$N_1^* = K_1 - N_2^* \frac{c_{22} + c_{21}w_{11}/w_{12}}{c_{12} + c_{11}w_{11}/w_{12}} \quad (\text{from Eq. 9}), \quad (11)$$

and

$$N_2^* = K_2 - N_1^* \frac{c_{12} + c_{11}w_{21}/w_{22}}{c_{22} + c_{21}w_{21}/w_{22}}, \quad (12)$$

where K_1 is the carrying capacity of species 1 and K_2 is the carrying capacity of species 2, and K_1 and K_2 are defined by the resource model as

$$K_1 = \frac{D(S_2 + S_1w_{11}/w_{12} - B_1)}{c_{12} + w_{11}c_{11}/w_{12}} \quad (13)$$

and

$$K_2 = \frac{D(S_2 + S_1w_{21}/w_{22} - B_2)}{c_{22} + w_{21}c_{21}/w_{22}}. \quad (14)$$

Note that Equations 11 and 12 are in the same form as Equations 1 and 2. This means that, at equilibrium, the model of competition for two perfectly substitutable resources which are supplied in an equable mode predicts that the competition coefficients will be

$$\alpha = \frac{c_{22} + c_{21}w_{11}/w_{12}}{c_{12} + c_{11}w_{11}/w_{12}} \quad (15)$$

and

$$\beta = \frac{c_{12} + c_{11}w_{21}/w_{22}}{c_{22} + c_{21}w_{21}/w_{22}}. \tag{16}$$

These values of α and β are predicted to be constant, independent of position along the resource ratio gradient. Thus, for perfectly substitutable resources supplied in an equable manner, the competition coefficients do not change along a resource ratio gradient, but the carrying capacities do (see Eqs. 13 and 14).

The same methods used in deriving this result were applied to the other cases presented in Chapter 7. MacArthur (1972) provides detailed mathematics for his derivation of the relationships between a model of competition for perfectly substitutable resources supplied in a logistic manner and the Lotka-Volterra equations. The formulae he derived were used to construct Figures 75 and 76. For essential resources, the relationship between the resource-based model and the Lotka-Volterra model was determined using a method similar to that outlined above. However, there is one difficulty which must be allowed for in these calculations. As illustrated in Figure 77, each species has certain regions of the resource plane in which it is limited by one or the other resource. It is necessary to solve the sets of equations for each of these regions. An outline of the solution is given in the main text.

References

Abrams, P. 1975. Limiting similarity and the form of the competition coefficient. *Theor. Pop. Bio.* 8: 356–375.

Abrosov, N. S. 1975. Theoretical investigation of the mechanism of regulation of the species structure of a community of autotrophic organisms. *Ekologiya* 6: 5–14. (Translated in *Soviet Journal of Ecology*, 1975, pp. 491–497.

Al-Mufti, M., C. Sydes, S. Furness, J. Grime, and S. Band. 1977. A quantitative analysis of shoot phenology and dominance in herbaceous vegetation. *J. Ecol.* 65: 759–792.

Armstrong, R. A., and R. McGehee. 1980. Competitive exclusion. *Amer. Natur.* 115: 151–170.

Ashton, P. S. 1977. A contribution of rainforest research to evolutionary theory. *Ann. Missouri Bot. Gard.* 64:694–705.

Bakelaar, R., and E. Odum. 1978. Community and population level responses to fertilization in an old-field ecosystem. *Ecology* 59: 660–671.

Beadle, N.C.W. 1954. Soil phosphate and the delimitation of plant communities in eastern Australia. *Ecology* 35: 370–375.

Beadle, N.C.W. 1966. Soil phosphate and its role in molding segments of the Australian flora and vegetation, with special reference to xeromorphy and sclerophylly. *Ecology* 47: 992–1007.

Beals, E. W., and J. B. Cope. 1964. Vegetation and soils in an eastern Indiana woods. *Ecology* 45: 777–792.

Blasco, D. 1971. Composión y distribución del fitoplancton en la region del afloramiento de las costas peruanas. *Inves. Pesq.* 35: 61–112.

Bond, W. 1981. Alpha diversity of southern cape fynbos. In F. J. Kruger, D. T. Mitchell, and J.V.M. Jarvis, eds.

Mediterranean-Type Ecosystems: The Role of Nutrients, Springer-Verlag, in press.

Box, T. W. 1961. Relationships between plants and soils of four range plant communities in south Texas. *Ecology* 42: 794–810.

Braakhekke, W. G. 1980. On coexistence: a causal approach to diversity and stability in grassland vegetation. *Agricultural Research Reports* 902, Centre for Agrobiological Research, Wageningen, The Netherlands. 164 pp.

Bradshaw, A. 1969. An ecologist's viewpoint. In I. Rorison, ed., *Ecological Aspects of the Mineral Nutrition of Plants*, pp. 415–427, Blackwell, Oxford.

Bradshaw, A., R. Lodge, D. Jowett, and M. Chadwick. 1958. Experimental investigations into the mineral nutrition of several grass species. I. Calcium level. *J. Ecol.* 46: 749–757.

Bradshaw, A., M. Chadwick, D. Jowett, R. Lodge, and R. Snaydon. 1960a. Experimental investigations into the mineral nutrition of several grass species. III. Phosphate level. *J. Ecol.* 48: 631–637.

Bradshaw, A., R. Lodge, D. Jowett, and M. Chadwick. 1960b. Experimental investigations into the mineral nutrition of several grass species. II. pH and Calcium level. *J. Ecol.* 48: 143–150.

Bradshaw, A., M. Chadwick, D. Jowett, and R. Snaydon. 1964. Experimental investigations into the mineral nutrition of several grass species. IV. Nitrogen level. *J. Ecol.* 52: 665–676.

Brenchley, W. 1924. *Manuring of Grassland for Hay*. Rothamsted Monographs on Agricultural Science. Harpenden, U.K. 146 pp.

Brenchley, W., and K. Warington. 1958. *The Park Grass Plots at Rothamsted*. Rothamsted Expt. Stn., Harpenden, U.K. 144 pp.

Campbell, E. 1927. Wild legumes and soil fertility. *Ecology* 8: 480–483.

Clements, F. E. 1916. *Plant Succession.* Carnegie Inst. Washington Pub. 242. 512 pp.

Coaldrake, J. E., and K. P. Haydock. 1958. Soil phosphate and vegetal pattern in some natural communities of southeastern Queensland, Australia. *Ecology* 39:1–5.

Colinvaux, P. A. 1973. *Introduction to Ecology.* Wiley, New York. 621 pp.

Connell, J. 1978. Diversity in tropical rainforests and coral reefs. *Science* 199: 1302–1310.

Connell, J. H., and R. O. Slatyer. 1977. Mechanisms of succession in natural communities and their role in community stability and organization. *Amer. Natur.* 111: 1119–1144.

Cooper, W. S. 1913. The climax forest of Isle Royale, Lake Superior and its development. *Bot. Gaz.* 55: 1–44, 115–140, 189–235.

Cooper, W. S. 1939. A fourth expedition to Glacier Bay, Alaska. *Ecology* 20:130–155.

Covich, A. 1972. Ecological economics of seed consumption by *Peromyscus*—a graphical model of resource substitution. *Trans. Conn. Acad. Arts and Sci.* 44: 71–93.

Cowles, H. C. 1899. The ecological relations of the vegetation on the sand dunes of Lake Michigan. *Bot. Gaz.* 27: 95–117, 167–202, 281–308, 361–391.

Crocker, R. L., and J. Major. 1955. Soil development in relation to vegetation and surface age at Glacier Bay, Alaska. *J. Ecol.* 43: 427–448.

Darwin, C. 1859. *The Origin of Species by Means of Natural Selection.* (Reprinted by The Modern Library, Random House, New York.)

Dayton, P. K. 1971. Competition, disturbance, and community organization: the provision and subsequent utilization of space in a rocky intertidal community. *Ecol. Monogr.* 41: 351–389.

Denslow, J. S. 1980. Gap partitioning among tropical rainforest trees. *Biotropica Suppl.* 12: 47–55.

deWit, C. T. 1960. On competition. Agricultural Research Reports (Versl. landbouwk. Onderz.) 66.8, Wageningen, The Netherlands. 88 pp.

Dix. R., and F. Smeins. 1967. The prairie, meadow and marsh vegetation of Nelson County, North Dakota. *Can. J. Bot.* 45: 21–58.

Doyle, R. W. 1975. Upwelling, clone selection, and the characteristic shape of nutrient uptake curves. *Limnol. Oceanogr.* 20: 487–489.

Droop, M. R. 1974. The nutrient status of algal cells in continuous culture. *J. Mar. Biol. Assoc. U. K.* 54:825–855.

Drury, W. H., and I.C.T. Nisbet. 1973. Succession. *J. of the Arnold Arboretum* 54: 331–368.

Dugdale, R. 1972. Chemical oceanography and primary productivity in upwelling regions. *Geoforum* 11: 47–61.

Egler, F. E. 1954. Vegetation science concepts 1. Initial floristic composition, a factor in old-field vegetation development. *Vegetation* 4: 412–417.

Ellis, R. C. 1971. Growth of *Eucalyptus* seedlings on four different soils. *Aust. For.* 35: 107–118.

Fedorov, V. D., and N. G. Kustenko. 1972. Competition between marine planktonic diatoms in monoculture and mixed culture.

Fischer, A. 1960. Latitudinal variations in organic diversity. *Evol.* 14: 64–81.

Foote, L. E., and J. A Jackobs. 1966. Soil factors and the occurrence of partridge pea (*Cassia fasciculata* Michx.) in Illinois. *Ecology* 47: 968–974.

Fox, J. F. 1977. Alternation and coexistence of tree species. *Amer. Natur.* 111: 69–89.

Garten, C. T. 1978. Multivariate perspectives on the ecology of plant mineral element composition. *Amer. Natur.* 112: 533–544.

Garwood, N. C., D. P. Janos, and N. Brokaw. 1979. Earthquake-caused landslides: A major disturbance to tropical forests. *Science* 205:997–999.

Gause, G. F. 1934. *The Struggle for Existence.* Hafner, New York.

Gleason, H. A. 1917. The structure and development of the plant. *Association Bull. Torrey Club* 44: 463–481.

Gleason, H. A. 1927. Further views on the succession-concept. *Ecology* 8: 299–326.

Goldblatt, P. 1978. An analysis of the flora of Southern Africa: its characteristics, relationships and origins. *Ann. Missouri Bot. Gard.* 65: 369–436.

Grime, J. 1973. Control of species density in herbaceous vegetation. *J. Environ. Manage.* 1: 151–167.

Groves, R. H., and K. Keraitis. 1976. Survival and growth of seedlings of three sclerophyll species at high levels of phosphorus and nitrogen. *Australian Journal of Botany* 24:681–690.

Grubb, P. 1977. The maintenance of species richness in plant communities: the importance of the regeneration niche. *Biol. Rev.* 52: 107–145.

Guillard, R., and P. Kilham. 1977. The ecology of marine planktonic diatoms. In D. Werner, ed. *The Biology of Diatoms*, pp. 372–469, Blackwell, Oxford.

Hanawalt, R. B., and R. H. Whittaker. 1976. Altitudinally coordinated patterns of soils and vegetation in the San Jacinto Mountains, California. *Soil Science* 121: 114–124.

Hanawalt, R. B., and R. H. Whittaker. 1977a. Altitudinal patterns of Na, K, Ca, and Mg in soils and plants in the San Jacinto Mountains, California. *Soil Science* 123: 25–36.

Hanawalt, R. B., and R. H. Whittaker. 1977b. Altitudinal gradients of nutrient supply to plant roots in mountain soils. *Soil Science* 123: 85–96.

Hansen, S. R., and S. P. Hubbell. 1980. Single-nutrient microbial competition: qualitative agreement between experimental and theoretically forecast outcomes. *Science* 207: 1491–1493.

Harper, J. L. 1969. The role of predation in vegetational diversity. In *Diversity and Stability in Ecological Systems, Brookhaven Symposium in Biology* 22: 48–62.

Harper, J. L. 1977. *Population Biology of Plants.* Academic Press, London. 892 pp.

Heinrich, B. 1976a. The foraging specializations of individual bumble bees. *Ecol. Monogr.* 46: 105–128.

Heinrich, B. 1976b. Resource partitioning among some eusocial insects: bumblebees. *Ecology* 57: 874–889.

Heinrich, B. 1979. *Bumblebee Economics.* Harvard University Press 245 pp.

Holdridge, L., W. Grenke, W. Hatheway, T. Liang, and J. Tosi, Jr. 1971. *Forest Environments in Tropical Life Zones: A Pilot Study.* Pergamon Press, Oxford.

Holling, C. S. 1959. The components of predation as revealed by a study of small mammal predation of the European Pine Sawfly. *Canad. Entom.* 91: 293–320.

Holm, N. P. and D. Armstrong. 1981. Role of nutrient limitation and competition in controlling the populations of *Asterionella formosa* and *Microcystis aeruginosa* in semicontinuous culture. *Limnol. Oceanogr.* 26:622–635.

Horn, H. S. 1974. The ecology of secondary succession. *Annu. Rev. Ecol. Syst.* 5: 25–37.

Horn, H. S. 1975. Forest succession. *Sci. Amer.* 232: 90–98.

Hrbackova, M., and J. Hrbacek. 1978. The growth rate of *Daphnia pulex* and *Daphnia pulicaria* (Crustacea: Cladocera) at different food levels. *Vestnik Ceskoslovenske Spolecnosti Zoologicke* XLII: 115–127.

Hsu, S. B., S. P. Hubbell, and P. Waltman. 1977. A mathematical theory for single-nutrient competition in continuous cultures of microorganisms. *SIAM J. Appl. Math.* 32: 366–383.

Hubbell, S. P., and P. A. Werner. 1979. On measuring the intrinsic rate of increase of populations with heterogeneous life histories. *Amer. Natur.* 113: 277–293.

Hulburt, E. M. 1970. Competition for nutrients by marine phytoplankton in oceanic, coastal and estuarine regions. *Ecology* 51:475–484.

Hulburt, E. M. 1982. Format for phytoplankton productivity in Casco Bay, Maine, and in the southern Sargasso Sea. Quoted with permission of the author; ms in review.

Huston, M. 1979. A general hypothesis of species diversity. *Amer. Natur.* 113: 81–101.

Huston, M. 1980. Soil nutrients and tree species richness in Costa Rican forests. *J. Biogeo.* 7: 147–157.

Hutchinson, G. E. 1959. Homage to Santa Rosalia, or why are there so many kinds of animals? *Amer. Natur.* 93: 145–159.

Hutchinson, G. E. 1961. The paradox of the plankton. *Amer. Natur.* 95: 137–145.

Hutchinson, G. E. 1978. *An Introduction to Population Ecology.* Yale University Press. 260 pp.

Janzen, D. H., H. B. Juster, and E. A. Bell. 1977. Toxicity of secondary compounds to the seed-eating larvae of the bruchid beetle *Callosobruchus maculatus. Phytochem.* 16: 223–227.

Kilham, P. 1971. A hypothesis concerning silica and the freshwater planktonic diatoms. *Limnol. Oceanogr.* 16: 10–18.

Kilham, S. S. and P. Kilham. 1981. The importance of resource supply rates in determining phytoplankton community In D. G. Meyers and J. R. Strickler, eds. *Trophic Dynamics of Aquatic Ecosystems.* AAS Symposium, in preparation.

Kirchner, T. 1977. The effects of resource enrichment on the diversity of plants and arthropods in a shortgrass prairie. *Ecology* 58: 1334–1344.

Kopczynska, E. E. 1973. Spatial and temporal variations in the phytoplankton and associated environmental factors in the Grand River outlet and adjacent waters of Lake Michigan. Ph.D. thesis, University of Michigan, Ann Arbor. 487 pp.

Kruckeberg, A. R. 1954. The ecology of serpentine soils. III. Plant species in relation to serpentine soils. *Ecology* 35: 267–274.

Kruger, F. J., and H. C. Taylor. 1979. Plant species diversity in Cape Fynbos: gamma and delta diversity. *Vegetation* 41: 85–93.

Lappe, F. 1971. *Diet for a Small Planet.* Ballantine, New York. 301 pp.

Lawes, J., and J. Gilbert. 1880. Agricultural, botanical and chemical results of experiments on the mixed herbage of permanent grassland, conducted for many years in succession on the same land. I. *Phil. Trans. Roy. Soc.* 171: 189–416.

Lawes, J., J. Gilbert, and M. Masters. 1882. Agricultural, botanical, and chemical results of experiments on the mixed herbage of permanent meadow, conducted for more than twenty years on the same land. II. The botanical results. *Phil. Trans. Roy. Soc.* 173: 1181–1413.

Lawrence, D. B. 1958. Glaciers and vegetation in Southeastern Alaska. *Amer. Sci.* 46:89–122.

Leon, J., and D. Tumpson. 1975. Competition between two species for two complementary or substitutable resources. *J. Theor. Biol.* 50: 185–201.

Leslie, P. H. 1948. Some further notes on the use of matrices in population mathematics. *Biometrika* 35: 213–245.

Levin, B. R., F. M. Stewart, and L. Chao. 1977. Resource-limited growth, competition, and predation: a model and experimental studies with bacteria and bacteriophage. *Amer. Natur.* 111: 3–24.

Levin, S. A. 1970. Community equilibria and stability, and an extension of the competitive exclusion principle. *Amer. Natur.* 104: 413–423.

Levington, J. S. 1979. A theory of diversity equilibrium and morphological evolution. *Science* 204: 335–336.

Levins, R. 1968. *Evolution in Changing Environments*. Princeton University Press. 120 pp.

Levins, R. 1979. Coexistence in a variable environment. *Amer. Natur.* 114: 765–783.

Lewis, E. R. 1972. Delay-line models of population growth. *Ecology* 53: 797–807.

Lotka, A. J. 1924. *Elements of Physical Biology*. Williams and Wilkins, Inc. 465 pp.

Lubchenco, J. 1978. Plant species diversity in a marine intertidal community: importance of herbivore food preference and algal competitive abilities. *Amer. Natur.* 112: 23–39.

Lund, J., F. Mackereth, and C. Mortimer. 1963. Changes in the depth and time of certain chemical and physical conditions and of the standing crop of *Asterionella formosa* Hass. in the north basin of Windermere in 1947. *Phil. Trans. R. Soc. London Ser. B. Biol. Sci.* 246: 225–290.

Lynch, M. 1978. Complex interactions between natural co-exploiters—*Daphnia* and *Ceriodaphnia*. *Ecology* 59: 552–564.

MacArthur, R. H. 1969. Species packing, and what interspecies competition minimizes. *Proc. Nat. Acad. Sci.* 64: 1369–1371.

MacArthur, R. H. 1970. Species packing and competitive equilibrium for many species. *Theor. Pop. Biol.* 1: 1–11.

MacArthur, R. H. 1972. *Geographical Ecology*. Harper and Row, New York. 269 pp.

MacArthur, R., and R. Levins. 1964. Competition, habitat selection, and character displacement in a patchy environment. *Proc. Nat. Acad. Sci.* 51: 1207–1210.

MacArthur, R. H., and R. Levins. 1967. The limiting similarity, convergence, and divergence of coexisting species. *Amer. Natur.* 101: 377–385.

MacArthur, R. H., and E. O. Wilson. 1967. *The Theory of Island Biogeography*. Princeton University Press. 203 pp.

Maguire, B. 1973. Niche response structure and the analytical potentials of its relationship to the habitat. *Amer. Natur.* 107: 213–246.

May, R. M. 1973. *Stability and Complexity in Model Ecosystems*. Princeton University Press. 235 pp.

May, R. M. 1975. Patterns of species abundance and diversity, In M. Cody and J. Diamond, eds. *Ecology and Evolution of Communities*, pp. 81–119. Harvard University Press.

May, R. M. 1978. The evolution of ecological systems. *Sci. Amer.* 239: 160–175.

May, R. M., and MacArthur, R. H. 1972. Niche overlap as a function of environmental variability. *Proc. Nat. Acad. Sci.* 69: 1109–1113.

McColl, J. G. 1969. Soil-plant relationships in a Eucalyptus forest on the south coast of New South Wales. *Ecology* 50: 354–362.

McCormick, J. 1968. Succession. pp. 22–35, 131, 132 in VIA, 1. Student publication, Graduate School of Fine Arts, University of Pennsylvania, Philadelphia.

McKone, M. 1980. The effect of nutrients on competition among old field plants. Masters thesis, University of Minnesota. 76 pp.

Mellinger, M., and S. McNaughton. 1975. Structure and function of successional vascular plant communities in central New York. *Ecol. Monogr.* 34: 161–182.

Milton, W. 1934. The effect of controlled grazing and manuring on natural hill pastures. *Welsh J. Agric.* 10: 192–211.

Milton, W. 1935. A comparison of the composition of hill swards under controlled and free grazing conditions. *Welsh J. Agric.* 11: 126–132.

Milton, W. 1940. The effect of manuring, grazing and cutting on the yield, botanical and chemical composition of natural hill pastures. *J. Ecol.* 28: 326–356.

Milton, W. 1947. The yield, botanical and chemical composition of natural hill herbage under manuring, controlled grazing and hay conditions. I. Yield and botanical. *J. Ecol.* 35: 65–89.

Monod, J. 1950. La technique de culture continue; théorie et applications. *Ann. Inst. Pasteur* 79: 390–410.

Moore, C. 1959. Interaction of species and soil in relation to the distributions of eucalypts. *Ecology* 40: 734–735.

Moore, C., K. Keraitis, and R. Groves. 1973. Effects of nitrogen on growth of *Eucalyptus gumminfera*, *E. agglomerata*, and *E. macrorhyncha* in sand culture. *Fld. Stn. Rec. Div. Pl. Ind. CSIRO* (Austr.) 12: 17–23.

Murdoch, W. W. 1969. Switching in general predators: Experiments on predator specificity and stability of prey populations. *Ecol. Monogr.* 39: 335–354.

Murdoch, W. W. 1971. The developmental response of predators to changes in prey density. *Ecology* 52: 132–137.

Murdoch, W. W., and J. R. Marks. 1973. Predation by coccinellid beetles: experiments on switching. *Ecology* 54: 160–167.

Murdoch, W. W., S. Avery, and M.E.B. Smyth. 1975. Switching in predatory fish. *Ecology* 56: 1094–1105.

Neill, W. E. 1975. Experimental studies of microcrustacean competition, community composition and efficiency of resource utilization. *Ecology* 56: 809–826.

Nelson, D., and J. Goering. 1978. Assimilation of silicic acid by phytoplankton in the Baja California and northwest Africa upwelling systems. *Limnol. Oceanogr.* 23: 508–517.

O'Brien, W. J. 1974. The dynamics of nutrient limitation of phytoplankton algae: a model reconsidered. *Ecology* 50: 930–938.

Odum, E. P. 1969. The strategy of ecosystem development. *Science* 164: 262–270.

Paine, R. T. 1966. Food web complexity and species diversity. *Amer. Natur.* 100: 65–75.

Paine, R. T. 1969. A note on trophic complexity and community stability. *Amer. Natur.* 103: 91–93.

Patrick, R. 1963. The structure of diatom communities under varying ecological conditions. *Ann. N.Y. Acad. Sci.* 108: 359–365.

Patrick, R. 1967. The effect of varying amounts and ratios of nitrogen and phosphate on algae blooms. *Proc. 21st Ann. Industrial Waste Conference*, pp. 41–51, Purdue Univ., Engineering Extension Dept. Bulletin 121, Lafayette, Indiana.

Patten, B. 1962. Species diversity in net phytoplankton of Raritan Bay. *J. Mar. Res.* 20: 57–75.

Pearsall, W. H. 1930. Phytoplankton in the English Lakes. I. The proportions in the waters of some dissolved substances of biological importance. *J. Ecol.* 18: 306–325.

Pearsall, W. H. 1932. Phytoplankton in the English Lakes. II. The composition of the phytoplankton in relation to dissolved substances. *J. Ecol.* 20: 241–262.

Petersen, R. 1975. The paradox of the plankton: an equilibrium hypothesis. *Amer. Natur.* 109: 35–49.

Phares, R. 1971. Growth of red oak (*Quercus rubra* L.) seedlings in relation to light and nutrients. *Ecology* 52: 669–672.

Pielou, E. C. 1969. *An Introduction to Mathematical Ecology*. Wiley-Interscience, New York. 286 pp.

Pigott, C. D., and K. Taylor. 1964. The distribution of some woodland herbs in relation to the supply of nitrogen and phosphorus in the soil. *J. Ecol.* 52 (suppl): 175–185.

Platt, W., and I. Weis. 1977. Resource partitioning and competition within a guild of fugitive prairie plants. *Amer. Natur.* 111: 479–513.

Poore, M.E.D. 1968. Studies in Malaysian rain forest. *J. Ecol.* 56: 143–196.

Porter, K. G. 1973. Selective grazing and differential digestion of algae by zooplankton. *Nature* 244: 179–180.

Porter, K. G. 1976. Enhancement of algal growth and productivity by grazing zooplankton. *Science* 192: 1332–1334.

Powers, C., D. Schults, K. Malueg, R. Brice, and M. Schuldt. 1972. Algal responses to nutrient additions in natural waters. II. Field experiments. In G. Likens, ed. *Nutrients and Eutrophication*, pp. 141–156, Am. Soc. Limnol. Oceanogr., Lawrence, Kansas.

Preston, F. 1948. The commonness, and rarity, of species. *Ecology* 29: 254–283.

Rapport, D. J. 1971. An optimization model of food selection. *Amer. Natur.* 105: 575–587.

Rapport, D. J., and J. E. Turner. 1975. Feeding rates and population growth. *Ecology* 56: 942–949.

Rapport, D. J., and J. E. Turner. 1977. Economic models in ecology. *Science* 195: 367–373.

Rhee, G-Y. 1973. A continuous culture study of phosphate uptake, growth rate and polyphosphate in *Scenedesmus* sp. *J. Phycol.* 9: 495–506.

Rhee, G-Y. 1974. Phosphate uptake under nitrate limitation by *Scenedesmus* sp. and its ecological implications. *J. Phycol.* 10: 470–475.

Rhee, G-Y. 1978. Effects of N:P atomic ratios and nitrate limitation on algal growth, cell composition, and nitrate uptake. *Limnol. Oceanogr.* 23: 10–25.

Rhee, G-Y. and I. J. Gotham. 1980. Optimum N:P ratios and the coexistence of planktonic algae. *J. Phycol.* 16:486–489.

Ricklefs, R. E. 1977. Environmental heterogeneity and plant species diversity: a hypothesis. *Amer. Natur.* 111: 376–381.

Riebesell, J. 1974. Paradox of enrichment in competitive systems. *Ecology* 55: 183–187.

Rorison, I. 1968. The response to phosphorus of some ecologically distinct plant species. I. Growth rates and phosphorus absorption. *New Phytol.* 67: 913–923.

Rosenzweig, M. 1971. Paradox of enrichment: destabilization of exploitation ecosystems in ecological time. *Science* 171: 385–387.

Russell-Hunter, W. 1970. *Aquatic Productivity*. Macmillan, New York. 306 pp.

Salisbury, F., and C. Ross. 1969. *Plant Physiology*. Wadsworth, Belmont. California. 747 pp.

Schelske, C. L. 1975. Silica and nitrate depletion as related to the rate of eutrophication in Lakes Michigan, Huron, and Superior. In A. D. Hasler, ed. *Ecological Studies*, vol. 10, pp. 277–298. Springer-Verlag.

Schelske, C. L., and E. Stoermer. 1971. Eutrophication, silica depletion, and predicted changes in algal quality in Lake Michigan. *Science* 173: 423–442.

Schindler, D. 1977. Evolution of phosphorus limitation in lakes. *Science* 195: 260–262.

Shields, L. M. 1957. Algal and lichen floras in relation to nitrogen content of certain volcanic and arid range soils. *Ecology* 38: 661–663.

Silvertown, J. 1980. The dynamics of a grassland ecosystem: Botanical equilibrium in the park grass experiment. *J. of Applied Ecology* 17: 491–504.

Smayda, T. 1975. Net phytoplankton and the greater than 20 micron phytoplankton size fraction in upwelling waters off Baja California. *Fish. Bull.* 73: 38–50.

Smith-Gill, S. J., and D. E. Gill. 1978. Curvilinearities in the competition equations: an experiment with ranid tadpoles. *Amer. Natur.* 112: 557–570.

Snaydon, R. W. 1962. Micro-distribution of *Trifolium repens* L. and its relation soil factors. *J. Ecol.* 50: 133–143.

Specht, R. L. 1963. Dark Island heath (Ninety Mile Plain, South Australia). VII. The effect of fertilizers on composition and growth, 1950–60. *Australian Journal of Botany* 11: 62–66.

Specht, R. L. and P. Rayson. 1957. Dark Island heath (Ninety Mile Plain, South Australia.) I. Definition of the ecosystem. *Australian Journal of Botany* 5: 52–85.

Sprugel, D. G., and F. H. Bormann. 1981. Natural disturbance and the steady state in high-altitude balsam fir forests. *Science* 211: 390–393.

Steel, R. G. D., and J. H. Torrie. 1960. *Principles and Procedures of Statistics.* McGraw-Hill, New York. 481 pp.

Steeman-Nielsen, E. 1954. On organic production in the oceans. *J. Inter. Council Stud. Conf. Sea* 19: 309–328.

Stewart, F. M., and B. R. Levin. 1973. Partitioning of resources and the outcome of interspecific competition: a model and some general considerations. *Amer. Natur.* 107: 171–198.

Stoermer, E. F., B. G. Ladewski, and C. L. Schelske. 1978. Population responses of Lake Michigan phytoplankton to nitrogen and phosphorus enrichment. *Hydrobiologia* 57: 249–265.

Taylor, P., and P. Williams. 1975. Theoretical studies on the coexistence of competing species under continuous flow conditions. *Can. J. Microbiol.* 21: 90–98.

Thurston, J. 1969. The effect of liming and fertilizers on the botanical composition of permanent grassland, and on the yield of hay. In I. Rorison, ed. *Ecological Aspects of the Mineral Nutrition of Plants*, pp. 3–10, Blackwell, Oxford.

Tilman, D. 1977. Resource competition between planktonic algae: an experimental and theoretical approach. *Ecology* 58: 338–348.

Tilman, D. 1980. Resources: a graphical-mechanistic approach to competition and predation. *Amer. Natur.* 116: 362–393.

Tilman, D. 1981. Experimental tests of resource competition theory using four species of Lake Michigan algae. *Ecology* 62: 802–815.

Tilman, D., and S. Kilham. 1976. Phosphate and silicate growth and uptake kinetics of the diatoms *Asterionella formosa* and *Cyclotella meneghiniana* in batch and semicontinuous culture. *J. Phycol.* 12: 375–383.

Tilman, D., M. Mattson, and S. Langer. 1981. Competition and nutrient kinetics along a temperature gradient: an experimental test of a mechanistic approach to niche theory. *Limnol. Oceanogr.* 26: 1020–1033.

Titman, D. 1976. Ecological competition between algae: experimental confirmation of resource-based competition theory. *Science* 192: 463–465.

Turelli, M. 1978. Does environmental variability limit niche overlap? *Proc. Nat. Acad. Sci.* 75: 5085–5089.

Van den Bergh, J. P. 1969. Distribution of pasture plants in relation to chemical properties of the soil. In I. Rorison, ed. *Ecological Aspects of the Mineral Nutrition of Plants*, pp. 11–23, Blackwell, Oxford.

Van den Bergh, J. P. and W. T. Elberse. 1975. Degree of interference between species in complicated mixtures. XII International Botanical Congress, Leningrad. *Abstracts* 1: 138.

Van den Bergh, J. P. and W. G. Braakhekke. 1978. Coexistence of plant species by niche differentiation. In A.H.J. Freysen and J. W. Woldendorp, eds., *Structure and Functioning of Plant Populations.* North Holland Publishing Co., Amsterdam, pp. 125–138.

Vandermeer, J. H. 1975. On the construction of the population projection matrix for a population grouped in unequal stages. *Biometrics* 31: 239–242.

Volterra, V. 1931. Variations and fluctuations of the number of individuals in animal species living together. In R. N. Chapman, ed. *Animal Ecology*, pp. 409–448, McGraw-Hill, New York.

Wagle, R. F., and J. Vlamis. 1961. Nutrient deficiencies in two bitterbrush soils. *Ecology* 42: 745–752.

Warren, R. G., and A. E. Johnston. 1963. *The Park Grass Experiment.* Rothamsted Report for 1963, pp. 240–262. Rothamsted Experimental Station, Harpenden, U.K.

Werner, P. A., and W. J. Platt. 1976. Ecological relationship of co-occurring goldenrods (*Solidago*: Compositae). *Amer. Natur.* 110: 959–971.

Whiteside, M. C., and R. V. Harmsworth. 1967. Species diversity in Chydorid (Cladocera) communities. *Ecology* 48: 664–667.

Whittaker, R. H. 1965. Dominance and diversity in land plant communities. *Science* 147: 250–260.

Whittaker, R. H. 1975. *Communities and Ecosystems*, Macmillan, New York. 385 pp.

Whittaker, R. H., and W. A. Niering. 1975. Vegetation of the Santa Catalina mountains, Arizona. V. Biomass, production, and diversity along the elevation gradient. *Ecology* 56: 771–790.

Williams, L. 1964. Possible relationships between plankton-diatom species numbers and water-quality estimates. *Ecology* 45: 809–823.

Willis, A. 1963. Braunton Burrows: the effects on the vegetation of the addition of mineral nutrients to the dune soils. *J. Ecol.* 51: 353–374.

Willis, A., and E. Yemm. 1961. Braunton Burrows: mineral nutrient status of the dune soils. *J. Ecol.* 49: 377–390.

Woodwell, G. M., R. H. Whittaker, and R. A. Houghton. 1975. Nutrient concentrations in plants in the Brookhaven oak-pine forest. *Ecology* 56: 318–332.

Yoshiyama, R. M., and J. Roughgarden. 1977. Species packing in two dimensions. *Amer. Natur.* 111: 107–121.

Young, V. 1934. Plant distribution as influenced by soil heterogeneity in Cranberry Lake region of the Adirondack Mountains. *Ecology* 15: 154–196.

Zedler, J., and P. Zedler. 1969. Association of species and their relationship to microtopography within old fields. *Ecology* 50: 432–442.

Zevenboom, W., J. Van Der Does, K. Bruning, and L. B. Mur. 1981. A non-heterocystous mutant of *Aphanizomenon flos-aquae*, selected by competition in light-limited continuous culture. *FEMS Microbiology Letters* 10: 11–16.

Author Index

Subject Index

LIBRARY OF CONGRESS CATALOGING IN PUBLICATION DATA

Tilman, David, 1949–
Resource competition and community structure.

(Monographs in population biology; 17)
Includes bibliographical references and indexes.
1. Competition (Biology) 2. Biotic communities.
I. Title. II. Series.
QH546.3.T54 574.5 81-47954
ISBN 0-691-08301-0 AACR2
ISBN 0-691-08302-9 (pbk.)